Designing America's
Waste Landscapes

Center Books on Contemporary Landscape Design

Frederick R. Steiner   George F. Thompson
Consulting Editor   *Series Founder and Director*

Published in cooperation with the Center for American Places,
Santa Fe, New Mexico, and Staunton, Virginia

Designing America's Waste Landscapes

Mira Engler

The Johns Hopkins University Press
Baltimore and London

© 2004 The Johns Hopkins University Press
All rights reserved. Published 2004
Printed in the United States of America on acid-free paper
9 8 7 6 5 4 3 2 1

The Johns Hopkins University Press
2715 North Charles Street
Baltimore, Maryland 21218-4363
www.press.jhu.edu

Library of Congress Cataloging-in-Publication Data

Engler, Mira, 1957–
    Designing America's waste landscapes / Mira Engler.
        p.   cm. — (Center books on contemporary landscape design)
"Published in cooperation with the Center for American Places, Santa Fe,
New Mexico, and Staunton, Virginia."
Includes bibliographical references and index.
    ISBN 0-8018-7803-9 (alk. paper)
    1. Waste disposal sites—United States.   2. Landscape design—United
States.   I. Center for American Places.   II. Title.   III. Series.
    TD788.E54 2004
    711'.8—dc21

                                                              2003010634

A catalog record for this book is available from the British Library.

To my mother Lea Sher Engler
and to my father Asher Engler

"the Pioneer of cleanliness . . .
the first artisan of state,
the billeting officer of culture"

Adolf Loos

Contents

Illustrations appear following pages 74 and 174.

Preface

About ten years ago, my colleague Gina Crandell and I met to brainstorm about a possible theme for a proposal for the Once and Future Park competition held by the Walker Art Center and the Minneapolis College of Art and Design. A desire to use wasted neighborhood open space, such as back alleys and small parks, and rising concerns about waste management converged in our entry, "Open Waste System Parks," which won an award in the competition. Since then, I have immersed myself in a research about waste and society and waste landscapes. Though the subjects are seemingly hideous and prosaic, I found them intriguing, mysterious, and paradoxical, symbols encumbered by deep-seated social and aesthetic taboos. I felt an urge to open up waste landscapes and set free their contradictory powers.

Over the years I have been able to discern the seeds planted early in my childhood that shaped my interest and landscape aesthetics and that found their way into this book. I grew up in a small duplex in the city of Holon, Israel. Our yard, which I called Dad's garden, was a small, utilitarian place of production. Vegetable beds below, fruit trees above, storage sheds in the corner, piles of compost and junk (copper, steel, tin) at the margins, and chickens and pigeons all found their respected place, use, and value; nothing was thrown away, and everything was reused and recycled. I grew up indulging in this garden. It was a sensible and frugal place, rich in forms, textures, and smells, a fertile ground for imagination and improvisation. It was a dialectical place—messy and orderly, dirty and clean—nowhere near beautiful by the usual aesthetic standards, which I could readily find embodied in the pristine and manicured English-style public garden nearby. The two gardens served as containers and backdrops for my childhood. But it was Dad's garden where I had my first lesson in landscape aesthetics.

My father's occupation as a plumber taught me another lesson. It brought me closer to the inner technological workings of our modern world. In those days, neither I nor anyone else I knew thought of a plumber as the pioneer of cleanliness who brought civilization to the world—in the words of the late-nineteenth-century Viennese architect and critic Adolf Loos, the "billeting officer of culture" (1997, 18). I knew only that my father was called on to fix and install pipes, sinks, and toilets and to clean sewers. He engaged in the most repulsive, yet most intimate and necessary work. When we remodeled our house in the mid-1960s, my father refurbished the most breathtaking bathroom for our family, a spacious room with the newest imported fixtures: two sinks, a tub, a toilet, a shower, and a bidet (something our neighbors had not heard of before). People from all over town came to see the pristine temple of pleasure in which we proudly indulged.

As a landscape architect, I came to realize that people regard members of our profession as apostles of beauty, where "beauty" means parks and gardens filled with trees, flowers, and lawns and shaped in the image of benign and pretty nature. With this realization came a certain restlessness. I searched for the dialectics in the landscape and eventually found it in extreme, marginal landscapes, in garbage dumps and sewage plants. I consider these places to be rich with cultural significance and substance; they are landscapes of extremes and contradictions, at once dreadful and fascinating, attractive and repelling. I found inspiration in architect Bernard Tschumi's notion that boundaries, edges, margins are the best places to break new ground and ask meaningful questions that might not be posed from the center. I decided to reside in the margin. Looking for ways to protect marginal, waste places from a second destruction, from the thoughtless desires of planners and landscape architects to dismiss them as morbid, consider them a disturbance, and fix them into prescribed formulas of beauty, I set out to write this book. The book is, therefore, a critique of society and of my profession.

Several people were important in shaping and realizing this book. My friend and colleague Gina Crandell, with whom I have collaborated on various public art projects, showed me the power of writing. Mark Chidister, another friend and colleague, supported and encouraged my early investigations into waste places and helped shape the book proposal. Kenneth Helphand, Niall Kirkwood, and Heidi Hohmann provided helpful comments on the manuscript draft. Timothy Keller, the head of the Department of Landscape Architecture at Iowa State University, heartened me and accommodated my needs for the project. The department and the College of Design at Iowa State University have consistently supported my research and travel through mini-grants and a Faculty Improvement Leave. The Graham Foundation provided a significant grant in 2000. The Iowa State University Publication Endowment

Fund supported this project in 2003. Matt Tucker, Dorothy Tang, and Peter Butler served as dedicated research assistants and helped me greatly with library research and editing. In particular, Peter Butler extensively researched toxic and radioactive waste landscapes and cowrote parts of the epilogue. Editor Alice Falk greatly enhanced the final manuscript. And finally, Mierle Laderman Ukeles, the indisputable authority on waste and society in the art world, served as a model and inspiration. Thanks to all these people and organizations, this book has been realized.

An earlier version of chapter 2 appeared as "Repulsive Matter: Landscapes of Waste in the American Middle-Class Residential Domain," *Landscape Journal* 16, no. 1 (1997), and is used with permission of the University of Wisconsin Press.

Introduction

garbage has to be a poem of our time because
garbage is spiritual, believable enough

to get our attention, getting in the way, piling
up, stinking, turning brooks brownish and

creamy white: what else deflects us from the
error of our illusionary ways, not a temptation

to trashlessness, that is too far off, and,
anyway, unimaginable, unrealistic . . .

—A. R. Ammons, *Garbage* (1993)

In *Garbage,* A. R. Ammons rummages in the garbage and discloses its constructive powers to transform and rebuild us. He delves into the muck in order to free us from a doom-laden cultural discourse that would rivet us in guilt and anxiety, and he thereby gives us hope. Not unlike Ammons's poem, this book uses waste landscapes as a cultural mirror in which we may view ourselves and ponder where to go.

For reasons that presumably are quite obvious, objects of cultural scorn, such as garbage cans, privies, and sewers, have rarely been subjected to scholarly study. But it is precisely because of their hideous, mundane nature and what they symbolize that I have concerned myself with them. Earthly, repulsive places (and containers) used to store, process, and get rid of our solid and liquid wastes mirror the rejects of society and are cultural manifestations as important and as telling as pristine, treasured places.

Waste is a pervasive, essential, and constructive process in human society. By *waste* or *garbage,* I mean residue material, objects, and environments that have been rejected or damaged and deemed unuseful (although this work deals primarily with domestic and municipal, rather than industrial, waste). We need "places of waste" where rejected goods and materials can be circulated (Lynch 1990). It is the contention of this book that waste is a key element in the ways in which we order and shape our environments, and it is inseparably intertwined and evolving with those environments. Waste creates geography. Waste disposal requires setting boundaries and creating margins, both physical and symbolic (Douglas 1966). We place waste in marginal spaces for removal or further transactions. As boundaries for those spaces are redrawn and margins shift, so does the placement of waste (Strasser 1999).

This book dwells on the margins, the landscapes of waste. It concerns these private and public landscapes at various scales: from a house (basement, bathroom), to a yard (outbuilding, garbage can, compost pile), to a neighborhood (back alley), to a city (sewage plant, dump), to a region (garbage transfer station). Its primary focus, however, is on public waste institutions—dumps, recycling and reuse centers, and sewage facilities. It probes people's perception of waste and waste landscapes and discusses the evolution, condition, role, and design of waste places in relation to the matrix of the American culture and landscape. Public waste landscapes (and waste) therefore serve as grounds for a general critique of culture and design.

American agrarian society produced little waste, which was generally benign. Urbanization, industrialization, and codes tying civility to cleanliness turned waste into a problem. Cities became great sinks of stench and garbage and breeding grounds for epidemics. Over the past two centuries, processes to control and eliminate waste were set in place in the private and public domains. Waste was privatized while also being transferred underground to a public but hidden milieu. These significant developments both reflected and enabled modern cultural ideals of progress and cleanliness. Gradually, America turned its waste, ever-growing in volume and type—human and industrial, bodily, solid, liquid and gas, benign, toxic, and radioactive—into contemptible and harmful matter and declared war on it.

Public waste landscapes served as the battlegrounds. First, we released waste gases into the air; excreted liquid waste into rivers, lakes, and harbors; dumped solid waste into gorges and oceans; and buried toxic and radioactive waste in barrels. Then, when air turned dark, water foul, and land toxic, we set up institutions to contain, correct, and destroy our wastes—asylums, prisons, and mausoleums, from which they always managed to escape. For our solid wastes we built incinerators, but they choked us with their emissions. We allocated waste to designated dumps, but they released their toxins and reek back at us. We sanitized the dumps, rebuilt them with tighter scientific control, and called them sanitary landfills, but they kept excreting methane and leachate broth. We then raised their standards, expanded their size, and moved them as far away from us as we could, building transfer stations to ease and optimize the transport of garbage. For liquid wastes we devised an underground labyrinth of intertwining pipes and pumps, but it vomited back its fetid muck and smelled. We built plants to filter the liquids, to refine the sewage mechanically and chemically, but we ended up with piles of slimy sludge. When we dumped the sludge in the ocean, it washed onto our shores. When we sprayed it on land, toxicity surfaced. Then we built dewatering facilities to dry and cleanse the sludge. For our toxic and radioactive waste we built even grander and deeper catacombs, ticking time capsules. This book is about all these places that we reluctantly created to fight our wastes.[1]

Two key questions follow. The first concerns how to avoid the problems created by waste and the places that support such venues; the second involves ways to deal with already existing places of waste. Society considers waste a private matter but a public issue. Waste has been distanced and repressed, yet it shapes our lives and landscapes. I contend that waste should be brought closer to our everyday environments and normalized and systems of waste treatment should be decentralized, with aesthetics employed to facilitate the change. Society deems waste landscapes inherently inferior sites. I contest this preconceived, negative, and hierarchical assumption and explore how design and artworks can change these perceptions into positive and constructive ideas and mold them into new spatial possibilities. In fact, I find marginalized waste landscapes to be complementary to highly valued landscapes, and attractive and productive in their own way. Rejected, polluted, and chaotic, they constitute some of the most repulsive and dangerous yet intriguing and challenging cultural spaces, constituting a high-energy reservoir for social and ecological health. At times, their haunting, unpredictable, and dynamic nature makes them a contemporary wilderness, the sublime of the beginnings of the twenty-first century. At other times, they become contemporary public arenas and gardens, sources of information, points of connection to the environment, or centers of social programs and activism.

Waste, both matter and idea, is closely related to notions of "dirt" and "margin" (physically and conceptually). Waste is mostly dirty; dirt is often waste. The margin often includes waste; waste is always marginal. Ex-

pounding the concepts of marginality and dirt illuminates critical notions about waste. All three are associated with pollution. They are dynamic, elusive, and dialectical concepts, socially and aesthetically defined, and they are essential for social order. Their value, conferred by
society, is intrinsic to the classification and stratification of society, material culture, and landscapes into binary oppositions of high and low,
good and bad, valuable and unvalued that are hard to overcome. As
Michel Foucault observes, "Perhaps our life is still governed by a certain
number of oppositions that remain inviolable, that our institutions and
practices have not yet dared to break down[:] . . . oppositions that we regard as simple givens" (1986, 23). But postmodern thought and praxis
have seriously disputed and disrupted these oppositions, reclaiming some
of their seemingly adverse consequences as virtues. This book attempts
to make yet another modest contribution: to reexamine the nature of the
apparent oppositions between clean and dirty, between central and marginal landscapes (positively and negatively valued landscapes), to nurture
dialectical relationships between the margin and the center, and to focus
on the specificity and potential of waste, dirt, and marginalia.

Scholars in cultural criticism, politics, anthropology, sociology, economics, literature, philosophy, and geography have attempted to theo-
rize about the nature and workings of marginal cultures and landscapes.
Archaeologists, writers, poets, and photographers have sought to explore
and illuminate their flavors, stories, and mysteries. Artists and social activists have always burrowed into marginal waste landscapes, releasing
their subversive potential, shedding light, focusing attention, and undermining social and aesthetic sensibilities that have caused these places
to be seen as objects of cultural scorn. But landscape scholars have shied
away from and overlooked the aesthetic and design potential of waste
landscapes. And until very recently, many of the people who have had
the power to determine their fate—planners, environmental engineers,
architects, and landscape architects—were caught within the confines of
disciplinary and professional conventions, the bounds of social taboos,
and the cordons of ecological despair (and technological guilt). Their
practice was rooted mostly in the ideology of waste places as distur-
bances, problems that require a fix. They found themselves fully embracing the pragmatism of engineers and developers in an effort to clear
away these nuisances inexpensively and efficiently, to make places of
waste invisible by burying and masking them, or, alternately, to trans-

form them in accord with the prevailing canons of the "useful" and "beautiful."

Clearly, our discomfort with dirty and smelly environments triggers an impulse to tighten control over and sequester the working facilities and to tidy and cover up those that are closed. This, in turn, erases their inherent potential to teach us the lessons of survival implicit in the interactions between cultural / technological and natural processes and to facilitate the flow of used matter back into the matrix of culture and nature. It also undercuts their capacity to provide alternative experiences of public place and to enhance landscape heterogeneity and complexity. I thus advocate their protection from "destruction" and call for rescuing them from the thoughtless desires to "fix," co-opt, and beautify them by reducing, leveling, or hiding their differences and particularities.

In so doing, I do not overlook the pollution, nuisance, and stigma inflicted on these environments and their nearby inhabitants. Rather, I argue for the need to confront these issues with design and art that normalize and integrate places of waste into communal and public space in the everyday landscape. Such a challenge calls on us to use new technologies that not only address safety and health but also are conspicuously integral to the aesthetic and programmatic language of particular places. It charges us to design facilities that give back something of value to the community and become an asset and a source of pride. Finally, it summons us to harness the environmental regulations and costly technical considerations essential to constructing, managing, and reclaiming these landscapes so that we make them central to design solutions. Yet, this book is about neither technical issues pertaining to mitigation measures nor policy issues. It is not a "green" book focused on recycling and environmental advocacy. Trained as a landscape architect, a designer, and an artist, I am concerned with the wise use and cultural significance of waste as it is expressed and reflected in our daily lives and places of waste. How does society imagine these places and programs and give them form? What should be their relationship to urban space and public use? The premise is, therefore, that the conflict between waste and society rests largely on aesthetics and imagination.

Following the pioneering work of artists like Robert Smithson, a few environmental artists and designers and a handful of landscape scholars have probed the values and opportunities of marginal and waste landscapes. They engage in philosophical and design inquiries that free them to wander between disciplines and away from social conventions and to shape waste landscapes anew. These works initially opened waste facilities to public gaze and dissipated some of their stigma. More recently, waste landscapes have become environmentally safer, architecturally palatable, and more compatible with recreational use. Some have even turned into economically and socially viable places. Their dialectical nature is exposed and interrogated; their identities are reframed.

Dumps, the earthworks of rising mounds and filled-up sinks, are seen as sites laden with ample, contradictory imaginings and facts. They yield scientific data but are burdened with myths; they are built as monuments to last but host processes of entropy; they stand for progress but symbolize death; they conceal treasures and unsolved mysteries but disclose a sincere and potent picture of contemporary consumer culture; they harbor fears and guilt but offer hope. Some retired dumps (or their successors, sanitary landfills) are designed as cultural monuments and ritual sites that distill and sharpen our visions and provide us with high grounds from which to see ourselves, while others are envisioned as centers for processing material resources, for education, and for enterprise.

Recycling and reuse centers, the terminals for sorting and orchestrating the flow of used, cast-off matter and artifacts, are sites of multiple dimensions and transactions. They encompass material recovery procedures; economic and social exchange; and processes of decomposition, sorting, bargaining, and creative transformation; and they display spectacular scenes, oddities, and opposites (dirty and clean, chaotic and orderly, ugly and pretty). Consequently, some reuse and recycling places are designed to alter our aesthetic sensibility about recyclables and reusables, encouraging us to favor less regulated and sanitized spaces. Additionally, more places for the "life-cycling" of rejected things are being integrated into our everyday environment and are conceived as environmental centers, marketplaces, or the focus of social programs, as new "town commons" or "material resource theme parks."

Sewers or sewage treatment plants, our technological kidneys, the grand cleansing systems of our murky liquids, represent multivalent subjects. They repel and please, horrify and fascinate, hide and reveal, free and enslave; they serve as agents of cleanliness and progress and, at the same time, are loci of uncleanness and pollution; they create and support the proper, modern civilization, yet represent its other, subversive side; they are conceived as a monumental engineering feat, yet are shunned as inverse monuments to preserving health and urbanity. They reside at the interface of body and environment, private and public, technology and nature, and city and waterways. Accordingly, a number of sewage treatment and sludge processing plants are shaped as urban gateways to the environment, as information terminals about the urban and natural systems that they traverse and connect. Alternatively, some are shaped as nature reserves and wastewater gardens (wetlands and aquaculture), displaying the ingenuity of machine and nature and our place in it. Quite significantly, smaller-scale, decentralized systems (solar aquatic systems and Living Machines) are slowly being integrated into the everyday environment. Thus, waste landscapes' inherent ambiguities and dialectics make them very appealing as a subject of intellectual and aesthetic investigation.

This book consists of five chapters. The first chapter sets up the broader cultural and design frameworks of waste institutions / landscapes in theoretical and historical terms. It also assembles, organizes, and studies the myriad ideas and facts entangled in the social constructs of waste. It examines the related notions of dirt and margin, the dialectics of clean and dirty and of edge and center, and it tampers with deep-seated cultural taboos of waste and dirt (notions of entropy, decay, permanence, purity, and order) that underlie how we perceive and design waste places. Chapter 2 centers on the private domain, which serves as a locus to examine the ideas laid out in chapter 1. It sketches the evolution of waste spaces in the American urban and suburban private domain—house, yard, alley, and street—in light of the changing perceptions, needs, and technologies related to waste and hygiene.

Public landscapes of waste make up the following three chapters. Each is a thorough inquiry into one specific category of waste institutions—dumps in chapter 3, waste recycling and reuse centers in chapter 4, and sewage facilities in chapter 5. Each consists of three narratives. The first sheds light on the ambivalent and dialectical cultural meanings and roles of the particular waste landscape. A review of history and precedents, interwoven with the trajectory of artists' and designers' fascination and engagement with that institution, makes up the second narrative. It highlights the fulcrum of change, the shift that enabled the involvement of contemporary artists and designers in projects dealing with public waste landscapes. The third narrative compares and critiques prominent examples of waste-related design projects—speculative, built, and proposed—emphasizing merits and limits of each. These project descriptions are neither comprehensive nor mere factual. They stress design insights and the philosophical ground on which environmental planners, designers, and artists operate. They search for the eloquence, incisiveness, and symbolism invested in physical forms, as well as the congruities and disparities between intentions and results, investment and return, technology and pattern, content and form, site and context.

The discussion challenges current planning and design practice. It argues that waste landscapes' inherent idiosyncrasies and dialectics underscore society's tendency to demonize or victimize them. It also demonstrates the futility of applying to waste places traditional design and art concepts—concepts to which some designers cling stubbornly. It dwells on solutions that elicit and nurture a fresh, positive, and critical perspective of the meanings and value of places of waste for healthy relationships between culture and nature, and it advances exploratory design scenarios in which waste spawns productive socioeconomic landscapes.

The primary raw material for this study is the landscape itself. Visits to waste sites and conversations with the people who shaped them form the text. Much of the discussion is grounded in historical precedents and contexts, relying on material drawn primarily from secondary sources: books written by professional historians. The trenchant studies from which I have gained insights and the scholars to whom I am most indebted are *Purity and Danger* (1966) by the anthropologist Mary Douglas, *Wasting Away* (1990) by the planner and environmental designer Kevin Lynch, *Paris Sewers and Sewermen* (1991) by the social historian Donald Reid, *Waste and Want* (1999) by the cultural historian Susan Strasser, *History of Shit* (2000) by the psychoanalyst Dominique Laporte, and various writings of the urban technology historians Martin Melosi, Joel Tarr, and Stanley Schultz and the anthropologist and garbage archeologist William Rathje.

Waste Landscapes contends that rejected landscapes can tell us as much as cherished landscapes, that landscapes at the extremes, both loved and hated, are indicative of ourselves and exercise great power over us.[2] Dwelling in marginal waste landscapes allows and affords a continuous rereading of the complex relationships existing between culture and nature. Disturbing as they may be, marginal waste landscapes must be looked at anew, as sites charged with cultural conflicts, errors, and peculiarities, yet pregnant with possibilities. Our shame and fear of waste have made its facilities invisible, inaccessible, uncontrollable, and often unsafe. Instead of distancing people from waste, design can bring them closer to waste operations and help foster creative solutions to problems intrinsic to waste disposal, problems faced by all people. Designing waste landscapes should therefore become a potent act of recovery and illumination. This book provides models that illustrate such possibilities.

Designing America's
Waste Landscapes

CHAPTER 1

Contemplating Waste
Theories and Constructs

Waste Works

Nature's waste, culture's waste

Contrary to what many say, humans did not invent waste; nature did. The squirrel eats a nut and dumps the shell. In some places the ocean floor leaks oil and pollutes marine habitats. Guano, the excrement of seafowl, kills plants and renders the birds' own habitat sterile. Lava and ash spewed from volcanoes can destroy whole ecosystems. As part of nature, humans are no exception to this pattern. We throw away what we do not use; we pollute, sterilize, and destroy. The difference, however, is that nature has mostly perfected its "waste management system." In due time and at varying rates, shells decompose and fertilize new growth; devastated sea and land turn into adaptive habitats. All forms of waste are eventually consumed, used, and recycled in a chain of matter and energy flow. But humans have persistently mismanaged their waste, creating new types at an increasing pace and in excessive quantities without establishing recovery mechanisms that enable their flow and circulation back into the cultural / natural systems.

The key to this disparity might be our cultural constructs of and taboos regarding waste. These have largely interfered with wise management, confusing real and perceived dangers (Lynch 1990). Culture has created its own "parasite" in garbage, the toxicity and sheer volume of which inflicts hazards and sucks the life and energy from whole regions. There are those who claim that the mindless inability to cope with waste and excess consumption may be an intrinsic feature of society, as intrinsic as the deliberate ability to innovate and produce. "The problem of garbage," the art critic Patricia Phillips tells us, "concerns the way that societies organize either a selective or synthetic vision of the world. Garbage, or more precisely an attitude about it, is something that appears to bring differing populations to common agreement. Everyone produces it, and no one wants anything to do with it other than take it 'out'" (1990, 12).

Society's visions of the world and attitudes toward waste have greatly changed over the years. So have the management and places of waste. Currently, people are realizing that garbage cannot be reduced to merely a pragmatic and efficient system of disposal. Efforts are thus directed at reducing and monitoring the production and accumulation of waste and facilitating its recirculation. Moreover, waste is no longer seen as an evil; it has its merits and productive possibilities, and it is a link in the continuous flow of matter and energy. This flow is not modeled, however, solely after natural systems, since the flow of material in society is more complex than in nature; the two are in fact intertwined. The waste humans produce is a critical constituent of interrelated social, political, economic, and ecological matrices and is dealt with more openly and effectively. These ideas and practices are poised to shape radically different waste institutions within our future landscapes.

Wasteless prospects

Emerging and prospective waste landscapes are beginning to be reshaped. Dumps have nothing to do with a permanent removal from society. Some new dumps are temporary storage spaces where types of waste are sorted and piled in separate megacontainers until the right method of treating and reusing each is found (fig. 1.1). Others are processing sites, *recycling dumps* where material enters on one side, undergoes compression and biogeochemical conversion, and exits on the other side as a usable material. Closed dumps are strip-mined, reopened for excavation of raw material. Alternatively, closed dumps become genetics research centers and laboratories where plant scientists test their ability to evolve new, extreme biotopes or experiment with plants to detoxify contaminants. Some toxic dumps become giant energy sculptures. They are covered with a flexible structural membrane that crawls over the pregnant breathing earthform and stretches like a living organ while collecting and channeling toxic gas for energy. In time the structure will be dismantled and reused in another place.

New facilities for the recirculation of used goods decentralize the current system and bridge the gap between supply and demand. In the inner city, neighborhood parks house a modest and accessible recyclables collection shed and yard, where residents bring materials to be recycled and where those who need them—the cart-people, children, and penniless artists—can retrieve items to get a refund or use them for creative projects (fig. 1.2). In suburbia, small recycling centers of simple but elegant construction, built around a large, green courtyard, provide convenient locations for everyday recyclable drop-offs as well as for community information, programs, and casual gatherings. Community-organized reuse facilities act as resource material centers, fostering social and economic transactions that blend communalism and environmental activism. *Surplus and reuse exchange centers* modeled after flea markets or junkyards are managed by private or nonprofit organizations. Some go beyond their recycling functions and become regional attractions—theme parks with refurbished rides, junk art, and junk food vendors. Other sites serve as social programmers, providing jobs for the poor and projects for teenagers, hosting community yard sales and events, sponsoring art workshops and contests with entries made of reusables, and launching outreach programs to teach home improvement, bicycle repair, gardening, and urban farming.

Large urban transfer stations are designed as architectural envelopes, adaptable enclosures containing offices and a garbage-processing area as well as a reusables store, a gallery, and a workshop space for items retrieved from the waste stream (fig. 1.3). Counties set aside areas for citizen drop-offs next to landfills or transfer stations and the next day offer the items for sale. Semiannual community pick-up days become citywide social happenings, as curbsides are filled with unwanted goods and neighbors search for lucky finds. In vacant lots and infill locations, small greenhouses host fish tanks that are fed with high-protein feed made from human waste and food scraps, and vegetable beds are fertilized with castings from vermiculture (worms fed on organic waste). Roadsides and urban vacant lots are turned into community gardens and urban farms, where organic waste is composted, flowers and crops are grown, and mulch is offered for free.

Some conventional sewage plants merge with *constructed wetlands* and aquaculture for secondary and tertiary cleansing, as well as for fish culture. There, selected plants, fish, snails, and microbes purify the wastewater before it enters streams and reservoirs, creating wildlife habitat and a slice of nature for urbanites. Laboratories testing sludge, water quality, and smell are integrated into sewage plants and also serve as community and learning centers. Elsewhere, small-scale enclosed marshes or *solar aquatic* systems grow fish and hydroponic gardens while cleansing wastewater and at the same time produce feed, crops, ornamental plants, and biogas. Community-run greenhouses built in urban empty lots and on rooftops operate *living machines* and grow produce and urban gardens. In densely populated cities, multistory structures purify effluent as it trickles down a skyscraper of plants, finding its way to the tap as clean, cool water. On some city blocks *sewage solar walls,* glass-capped, terraced planters that run between sidewalk and street, progressively filter and purify wastewater. Floating garden islands are cleansing wastewater lagoons and urban streams. In suburbia, residential gardens act as wastewater cleansing systems. Composting facilities in central neighborhood locations process human waste collected from electronic composting toilets. And in the country, tree farms are irrigated by wastewater from animal farms.

At the manufacturing level, new alchemists are laboring to perfect a wasteless world—building up materials from scratch, eliminating wasteful fabrication and production methods, and creating durable and smaller objects that will last longer and can be efficiently recovered at the end of their useful life. The closed-loop eco-industrial park is also perfecting its own wasteless environment: one industry uses the waste or output of another in a chain of use and reuse, and buildings and processes are conceived as ecostructures, machines attuned to the Earth's unique biogeochemical processes.

These evolving landscapes of waste reflect a cultural project that redefines sustainability and practicability of waste management within broader cultural terms. The project aims to normalize and decentralize waste systems and reassign smaller facilities to the everyday environment. It also empties waste infrastructure of its environmental moralism and symbolism and endows it with overt aesthetics that are true to its multilayered ecological, social, and economic functions. *Infrastructure is turned to suprastructure.* The new cultural project represents a departure not only from the modern facilities built early in the twentieth century but also from the facilities and programs of the 1970s and 1980s postmodern city (which lingered well into the 1990s). These changes reflect a vision for a city that differs dramatically from those of the utopic and objective modernist and the apocalyptic and subjective postmodernist.

Wasteful visions

The scientifically based modernist ideal envisioned a city that swiftly and efficiently evacuated its wastes to faraway facilities through hidden underground systems. In 1876, energized by the prospects of science and technology, physician Benjamin Ward Richardson designed an imaginary place called Hygeia, a city of health. It is a city built on the premise of sanitary progress, control, prediction, and measure. It displays no waste or dirt and keeps disease and death in their "proper or natural place in the scheme of life" (1876, 11). Everything from land use, to city form and streets, to building mass and height, to density, infrastructure, pavement, kitchens, bathrooms, bedrooms, and personal spaces are prescribed to maximize natural light, ventilation, and sanitation. Early morning scavengers collect the daily garbage and manure from the streets. No bars or places of ill-repute exist in the city. Factories and workshops are allocated to specialized fringe areas; so are hospitals, asylums, and nursing homes for the elderly. Sanitary works and sewage farms are built at the distant urban margins. An officially designated administrator watches over Hygeia's sanitary welfare.

Richardson's positivist, mechanistic utopia reflects the worldview of modernism. He writes, "We, scholars of modern thought, have the broader, and therefore more solemn and obligatory knowledge, that however many to-morrows may come, and whatever fate they may bring, we never die[,] . . . communicating motion to the expanse before us, and leaving the history we have made on the shore behind" (8–9). This was a promising vision for a civilized, better, and happier world. But the course of history took a different direction. Garbage was merely removed from the private domain and city center to the public urban margins, enabling urban economic prosperity while encouraging waste and severely polluting broader natural systems.

The disillusioned, postmodernist image of cities conveyed a total loss of control. In his prophetic book *The Waste Makers* (1960), Vance Packard condemns the philosophy of wasting, in which prosperity is linked to disposability. He sketches a disturbing image of a hyperconsumptive city he calls Cornucopia. Everything in this city is made of throwaway products; nothing is repaired past its second year of service. Clothes, appliances, cars, and refrigerators are disposable. Homes are made of papiermâché. Since disposability advances progress and feeds the economy, assembly lines in factories eject all surplus products via a back door directly into a dump as soon as supply exceeds demand. Every Wednesday a warship is filled with surplus items, sailed out to sea, and sunk. People shop in giant supermarkets, where conveniently located push buttons are activated by an electronic lifetime ticket that everyone gets at birth. Shopping carts, first emptied of outdated products into the supermarket receptacles, are quickly filled up with new merchandise.

Society's pervasive wastefulness and blind trust in technological fixes are also captured in the introduction of Louis Blumberg and Robert Gottlieb's *War on Waste* (1989). They propose an ingenious solution to the waste problem: a space disposal program. A special division within NASA will guide daily launches to the moon, which will last three hundred years, when the moon reaches full capacity. Since "moonfilling" is cost-effective as long as certain volumes of waste are guaranteed, people are encouraged to waste. Moonfill proponents counter critics who worry about potential harm to the moon environment by arguing that "a waste site, after all, is nothing more than the environmental equivalent of a moonscape" (xx).

Other cynical postmodernists sketched two opposing catastrophic scenarios. One sees a consumptive civilization incapable of managing its wastes, choked in its own excessive effluent, poisoned by its own refuse; the other imagines a society sanitizing itself to death, creating no waste yet eliminating excess along with pleasure. In *Wasting Away* (1990), Kevin Lynch laid out two extreme and contrasting nightmare cities called cacotopias. In the wasteful cacotopia, waste ribbons, mountains filled to capacity surround the cities, abut those of nearby settlements and consume any leftover space there is. Frontier guards from rivaling cities make sure neighboring cities do not tip refuse over the ridge. Once dumping space is exhausted, garbage is directed to oceans and canyons and injected into mining shafts; its liquids and gases cause the ground to subside. Land and water are contaminated. Toxic gas is ejected into space in bags that cover the sky, hiding the sun from the city below. Aerial sweepers clear the approach lanes for airplanes landing at major airports. Dereliction and neglect prevail everywhere. Buildings fall into ruins; vacant lots are overgrown and dumped on; fossil fuels and forests are exhausted. Death, accidents, and short life-spans are the rule. The rich live in protected settlements while the poor are left outside to choke (4–5).

The wasteless cacotopia is no better. There is no waste, no leftovers, no excess. All is reused. Air and water are clean. All food is eaten immediately following its preparation. There are no weeds, no useless animals such as dogs and cats, no rats, no mosquitoes. There is no empty space or building, no useless backyard. Weather is controlled, and all is standardized for maximum efficiency. There is no waste of time, no mental drifting, no psychological depression. There is no waste of sperm and ova; sexual intercourse is regulated to control population. No gossip or advertisements are allowed. There is no place for superfluity and extravagance, for celebrations or decorations. Accidents and wars do not exist.

Life is prolonged and death is rare. This is a predictable world, "a silent world, disturbed only by soft, precise, symbolic communications" (8).

With similar cynicism, Italian writer Italo Calvino writes in *Invisible Cities* (1974) of Leonia, an imaginary city that buries itself in its own garbage. Leonia renews itself daily, throwing away all of yesterday's commodities. Much as we discard yesterday's paper, Leonians cast off all they have used the day before, without exception—food, dinner plates, dinner tables, chairs, clothing, bed sheets, refrigerators, books, information, musical notes. The city generates and consumes new things each day, to be thrown out the next. Street cleaners and garbage trucks, its most precious technology, sweep loads of clean plastic bags from sidewalks and expel them outside the city. As a result, "A fortress of indestructible leftover surrounds Leonia . . . like a chain of mountains. . . . The greater its height grows, the more the danger of a landslide looms: a tin can, an old tire, an unraveled wine flask, if it rolls toward Leonia, is enough to bring with it an avalanche. . . . A cataclysm will flatten the sordid mountain range, canceling every trace of the metropolis always dressed in new clothes" (115–16).

Throughout history such imaginary scenarios have served as expressions of social thought, as guides and warnings. They spawned timely cultural responses, which in turn shaped the Western history of waste landscapes and of our cities.

The r/evolution of waste: A cultural chronological framework

Although waste and its institutions are primarily linked to modernity, their history predates the modern era. Beginning in the Renaissance, four broad chronological cultural projects concerning waste can be discerned—the premodern, modern, postmodern, and supermodern.

The strides of premodernity. Evidently, the premodern era did not have public waste facilities, nor did it have many of the wastes that the word conjures up for us today. Nevertheless, its engagement with waste should be recognized as a full-fledged cultural project, which paved the way for the modern era's attempts at waste control. By the sixteenth century bodily waste and plentiful filth were ubiquitous in cities (fig. 1.4). It was a time in which the boundary between public and private spheres was indistinguishable and society was divided between the noble few and the masses. With the construction of the "individual" and the rise of the middle class, configurations of the private realm gradually appeared. Refined manners, new concepts of order and cleanliness, and the privatization and sublimation of waste began shaping class hierarchies and de-lineating private and public spheres (Laporte 2000).

Although London issued decrees in the fourteenth century to clear all streets of swine, dung, dirt, and filth and so keep them clean, they had little effect. Paris streets, glutted with animal and human excrement, dotted with heaps of offal, and prone to waves of fetid odors had to wait until the sixteenth century for reform. In his black velvet–covered book *History of Shit* (2000), Dominique Laporte pins down the exact date and event that set in motion the process perfected three hundred years later in the modern project: the November 1539 law that targeted the filth-mired streets of Paris. The decree mandated the construction of cesspools and the storage of refuse, offal, putrefying matter, urine, and excrement inside the confines of private homes and their prompt disposal into streams. According to Laporte, this was the origin of the domestication of waste. And, although most front yards and streets showed no sign that the new rules were being followed, a changed discourse concerning promiscuity and the links between propriety and property began to emerge in France and elsewhere in the European landscape. As streets (and language) were gradually deodorized, sight was elevated as the re-fined sense, replacing smell. Sewage dumps and slaughterhouses were set at a distance, at "proper" fringe locations. Garbage and dirt quickly be-came signifiers of the lower classes.

In the chaotic and disease-ridden emerging American cities of the late eighteenth century, similar decrees and processes were set in motion. There, too, streets were soaked in refuse and cleanliness was associated with gentility and class, garbage with the poor and immigrants. In ur-ban well-to-do areas and in the country, residents relegated waste to basements and, especially, the yard, clearly demarcating the boundary between low and high, inside and outside, private and public. Although society had little to dispose of and reused much, poor sanitation and re-curring epidemics in crowded city centers eventually spawned a decisive scientific and technological response.

The plumbing of modernity. Pipes and plumbing technology proclaimed modernity. They enabled the ultimate domestication of waste and its transfer to the public domain. They assigned bodily functions to the pris-tine toilet and bathroom. Pipes and sewers led to the restructuring of cities, as they were cut through, dug open, and surgically retrofitted for the supply of essential water and the evacuation of wastewater. Relegated to the secrecy of the underground, public waterworks came onto the scene in the early nineteenth century and sewers a few decades later. London, Paris, and large American cities first brought water to residences at the turn of the century and introduced sewers in the 1850s.[1]

The municipal networks carried out the new laws of the state, making possible different norms of behavior and cleanliness, now discussed under the terms "hygiene" and "sanitation." People began adopting sanitary habits and embraced the bathroom and its sanitary fixtures, seeing in them an embodiment of privacy, cleanliness, individualism, and progress. The modern project affirmed and fixed the scientific mind, widening the schism between humankind and natural law. As modernity proceeded aggressively toward ever-greater purity and order, repressing all signs of the abject with the ingenuity of the engineer—mucky fluids were discharged directly into rivers, garbage was destined for sanitized and distant dumps—excessive waste became symptomatic of the failure to engage nature positively. Simultaneously, social and spatial hierarchies defining low and high spheres were affirmed and tightened.

While the modern utopia entrusted its happiness to hygiene (and consumption), Sigmund Freud warned us that the danger to civilization comes from repressing thoughts of waste, or in his words "unthinking waste." In *Civilization and Its Discontents,* Freud (1961) articulates the essential need of society for order and cleanliness, but he insists that it comes at a great price—loss of happiness. For Freud excrement, truth, and pleasure are closely related. The taming of anal pleasures—particularly the pleasure of excreting—in early childhood, their repression by strict social taboos, and their replacement with accepted traits of parsimony, order, and cleanliness are responsible for much of human misery and bursts of violence. Refuse waste is not a threat in and of itself; its repression or avoidance create fear and the conditions for refuse to become a danger. Freud tells us that confronting the taboos associated with waste is facing up to the truth, to dirt and ugliness, and it is an essential freeing act.

In the last quarter of the twentieth century, postmodernism set out to deliver what Freud had called for. It sought to undermine the bureaucracy of hygiene, which drained the instinctive, destructive pleasures we have in dirt and pollution (Lahiji and Friedman 1997, 53). It also responded to dangers being posed to the environment. The sanitary, engineered, out-of-sight institutions that came back to haunt us, oozing their toxins into air and waterways, hastened the new cultural project.

The posture of postmodernism. The postmodern project came along in the 1960s to deconstruct established systems and expose inherent dilemmas—on the one hand, affluence, conspicuous consumption, and sophisticated technology; on the other, fixed social hierarchies, dwindling resources, and environmental pollution. Environmental protection became a prime consideration in politics and practice. During this time, sewage facilities and sanitary landfills were subjected to increasingly stringent pollution control standards that necessitated costly upgrades and new facilities. These reforms planted the seeds for nonconsumptive, alternative ecotechnologies and spurred new research. While municipalities transferred responsibility for waste management to giant corporations, environmental advocacy groups forged nonprofit recycling organizations. The public awakened to the threats, resisting the placement of waste facilities in their backyards. Consequently, society had to consider waste facilities anew, rescued from their sequestration and opened up to the public gaze and to a variety of new experiments.

The new project was founded on several cultural discourses. One highly provocative discourse took place in the art world. Building on the early-twentieth-century art critique of consumer culture and affluence, artists of the 1960s engaged the subjects of pollution and garbage. Environmental art was gradually shaped as social environmental action, using ecology as a new medium of inquiry to forge awareness and abate pollution. Scatology—the study of excrement and bodily ejaculations— was a related topic of interest. This work subverted the norm and exposed a ubiquitous side of life, making public that which had long been considered private.

Another topic that pervaded and defined postmodern discourse and had potent implications for waste landscapes concerned the marginal or the "other," as postmodernism disrupted traditional hierarchies and the oppositional relationship between margin and center (as well as culture and nature, male and female). It freed what was seen as marginal from its inferior position relative to the center, and it won recognition for critical role of marginal places and ideas as an alternative and a complement to the center. In postmodernism, the margin was considered a milieu where newness emerges and where the critical distance necessary to clearly see the center is maintained. This discourse, which pertained first to the sociopolitical arena, was expanded to include landscapes. Waste landscapes, the epitome of marginal lands, were recognized as unique sites that have the potential to ground social and cultural introspection and critique.

The messages put forth by artists and other thinkers—urging us to think about and see waste and advocating the inherent capacity of waste places to shed light on and supplement the center—have to some extent shaped postmodern designs of waste sites. Many of the marginal waste landscapes built then and presented as timely and innovative were, in fact, still hostage to the confines of the modern, but others were built with the new ideas in mind and gave rise to a new formal language, new symbols, and new functions. As long-standing public spaces ceased functioning as meeting places, new public spaces emerged around infrastructure and the exchange of waste, used commodities, and recycling.

The promise of supermodernity. The evolving scenarios presented at the outset of the chapter beckon us toward a new threshold and point to a new destination in the quest to live at ease with waste and make it effectual. New social ideas that began infiltrating thought in the 1980s led to significant changes in the 1990s. Globalization, increased mobility, radical individualization, electronic communication, and hyperconsumerism spawned an interconnected pluralistic culture where technological ingenuity is rewarded and instantly disseminated. The new conditions of supermodernity altered the experience of space and time. The anthropologist Marc Augé suggests that the new era is characterized by growing spatial anonymity and lack of shared meaning. Augé (1995) argues that public space changed form a social meeting space to a regulated domain for individual action, and he links the changing of the nature of public space to three conditions: abundance of space (or what he calls non-space, the mere spheres of mobility and consumption, such as malls and airports), abundance of signs and information, and abundance of individualism.

Architecture, in turn, lacks specific relations to site and context and becomes a flexible shell for sophisticated electronic technologies and information transfer (Ibelings 1998, 62). Waste landscapes are also built as sophisticated containers, material transfer stations, and transaction places at a variety of scales and facilitate the flow of matter back into society. Monitored by advanced ecotechnologies, some facilities are conducive to local community and individual operation and responsibility. They emphasize the practical rather than the symbolic, the everyday rather than the unique, the decentralized rather than centralized. They are conceived by collaborative teams of engineers, scientists, designers, and artists who shape the aesthetic dignity of each place and earn it attention and respect.

The new waste landscapes also build on the ambiguities and dialectics of waste, its true source of power.

The Ambiguity of Waste

Waste language

The dialectical nature of waste is evident first in the language in which it is discussed. Seemingly negative connotations of waste ultimately harbor a positive dimension as well. "Waste" comes from Latin *vastus,* meaning "unoccupied" or "desolate." The plentiful dictionary definitions of both verb and noun include such terms as barren, empty, loss, decrease, decay, decline, damaged, defective, unwanted, and rejected—negative concepts that entail uselessness, negligence, or human failure. Originally, waste implied a physical space uninhabited (or sparsely inhabited) and uncultivated, a desert or a wilderness, usually at civilization's edge. According to the *Oxford English Dictionary* (*OED* 1989) it could also refer to a vast expanse of ocean or of snow-covered land. Apparently useless, such land nevertheless served an important role. In the Middle Ages most English villages had some land technically termed "waste," though this was normally a worthwhile space containing much of value, including game, pigs, the supply of fuel and timber, and the possibility of village expansion (Pollard 1997, 15). Later on, wasteland acquired an additional meaning of ravagement or devastation, such as mined or bombed land.

The noun "waste" can also apply more generally to an action or process, to matter, and to an idea. As an action, waste means a useless or needless expenditure (of work, energy, time, or extravagant consumption). The word also denotes an act of destruction or devastation or a process of "gradual loss or diminution resulting from use, wear and tear, and decay" (*OED* 1989). As matter, waste is the totality of human discards—rejected materials, and objects of no use. We usually group such waste by chemical composition, as organic or inorganic; by effect, as benign or hazardous; and by origin or cause, which can be categorized four ways. *Used materials* are no longer suitable or of utility, whether they wear out, no longer fit current needs or standards, have fallen out of fashion, or are somehow damaged. *Leftovers* are the residue after useful parts have been consumed (e.g., packaging, vegetable peels). Often, they are integrated into the product, as "planned waste." Such residue is also one outcome of cleaning and sorting. *By-products* (or spents) are materials inevitably formed in producing something else (e.g., industrial tails); sometimes they are poisonous. By-products can also be merely surplus, thus unnecessary and ultimately discarded. Finally, *human waste* consists of bodily excretions and ejaculations. Waste professionals use other categories: household and commercial (municipal solid waste, or MSW), construction/demolition, and industrial. As an idea, waste is something of no or negative value, a value conferred by society. These judgments help order the physical and social world in hierarchical economic and moral systems. Such combinations as "waste of time" and "waste of words" imply something profitless, serving no purpose, superfluous, and in vain.

Cultural discourse, which has always controlled the values we give to words, legitimized the term "waste" over other terms, such as "garbage." Apparently, waste sounds genteel, vague, and inclusive, whereas garbage sounds derogatory, provocative, and contaminated. In the *Oxford English Dictionary*, "garbage" gets only a fraction of the space allocated to "waste." It appeared three hundred years later and was used more specifically and less extensively. "Garbage" was first applied to the offal of an animal prepared for food, then to the refuse or by-product of any process (usually organic). Interestingly, in an abstract sense it means "worthless and foul literary matter," and the citations for that meaning include one from 1882: "Any garbage is food for woman's vanity." Especially since the 1970s, "waste" has appeared more often in public discourse, replacing and encompassing other terms as garbage, refuse, rubbish, trash, junk, and litter.[2]

Seemingly, the language of waste does matter, yet it has always remained compounded, encumbered by a web of concepts and sentiments.

Waste matters

Kevin Lynch tells us that social symbols associated with the cleaning and disposal of waste can be "so deep-seated, or so closely linked to other social concepts, that it is disturbing to tamper with them" (1990, 39). According to Lynch, our discomfort with waste or dirt is as much a creation of our minds as a result of actual dangers.

Waste is inevitable. Making waste is both conscious and unconscious. House cleaning involves a conscious removal of dust, dirt, and unwanted detritus. But for many the disposal of refuse is one of our most unconscious acts. As William Rathje and Cullen Murphy tell us in *Rubbish! The Archaeology of Garbage*, "The cliché about garbage we've all heard is: 'out of sight, out of mind.' Yet even when it's *in* sight garbage somehow manages to remain out of mind" (1992, 45). Garbage, our own daily creation, passes under our eyes virtually unnoticed. It is placed in its specified container and the next day it disappears. Its continual turnover inhibits perception. (On the other hand, Americans are exquisitely sensitized to the existence of litter.) Despite, and maybe because of this obliviousness, all agree that waste never lies.

Waste is everyone's matter, and it is no one's. It is a private matter that becomes a public matter (as soon as it is put on the curb for collection or flushed down the drain). It is a private problem deferred to the public, ballooning from a relatively minor and simple undertaking for the individual to a major and complex one for collective society. The garbage problem is, to borrow a phrase from Garrett Hardin, a classic "tragedy of the commons" (1977, 16). Hardin uses the town commons— the shared pasture traditionally found in the center of early American and English towns—to explicate the concept of resource limits and the relation between private and public upkeep. The commons system of grazing works successfully as long as the number of beasts is below the supporting capacity of the land. But it is in the interest of each herdsman to increase the size of his own herd, even though when all do so, the limits of the commons are reached. They all race toward a clearly visible ruin. The behavior of those producing garbage is similar. Chemicals, detergents, fertilizers, batteries, toxic paint, and grounded organics are conveniently drained and discharged to common dumps and sewage plants. Private responsibility is replaced with public complacency and inability. Waste and its processing sites have turned into burden and havoc.

Waste is society's dirty secret world; so is sex. Several writers have drawn the analogy between garbage and sex. Everyone produces or does it (and can enjoy it), rich and poor. It must remain private, and when in public it becomes a dirty matter. In *Underworld* (1997), Don DeLillo weaves the business of waste and sex in a captivating episode in a desert hotel. A dignified group of waste management businesses assembling for their annual convention shares a seminar space with forty married couples there to trade sexual partners and openly talk about their feelings. Both parties become self-conscious about both matters and their inextricable relations.

Waste is ambiguous, elusive, and relative. Waste varies in its relation to people, context, class, culture, and time. There is no absolute waste matter. As the old saying goes, one person's trash is another's treasure. Manure is a valuable substance on the farm and a scorn in the city. Empty cans are junk for the wealthy and a source of living for the poor. Used glass jars are valuable commodities in Russian bazaars but treated as trash in America. The production of waste is also in a constant flux: over time, some material wastes are created, others are eliminated. Ashes were a common waste in the past but are no longer produced in bulk; conversely, paper, plastic, and yard waste are relatively new. What counts as waste changes over time, losing or gaining value as it moves into or out of the category of rubbish.

The social scientist Michael Thompson reminds us in *Rubbish Theory* (1979) that all cultures insist on some distinction between valued and valueless material culture. All objects fall into one of three categories: valuable, valueless, and negatively valued. Durable objects and places are valuable, the transitory are valueless, and the negatively valued are rubbish. An object itself has no intrinsic value, though value, once bestowed, is commonly self-perpetuating. According to Thompson, people near the top of the social scale hold the power of bestowing or denying value to things. They thereby ensure that their own objects and places are always durable and valuable and that those of others are always transitory and valueless. But change does take place. The boundaries between material waste and nonwaste are not fixed but move in response to social pressure and technological advances. Likewise, watches, which used to be durable items passed on from one generation to the next, are now flimsy and of little value. Conversely, Stevensgraphs (i.e., silk woven pictures sold as cheap souvenirs between the 1870s and 1940s), which had no market until the early 1960s, are now expensive collectible items (Thompson 1979). Such is also the case with physical space. Places of waste are not doomed permanently due to intrinsic physical properties. Unpopular and devalued places can become popular and valued and vice-versa. Their qualitative features are the result of social valuation, not the cause.

Waste is inherently dialectical. It is repulsive, and it brings joy. On the one hand, waste is encumbered by misconceptions and carries a tenacious stigma. It is accompanied by feelings of disgust and perceived as an offense and an obstacle, as something diseased that must be eliminated. On the other hand, creating and releasing waste and excess can be a source of human pleasure. The artist Robert Smithson explains, "There's a certain kind of pleasure principle that comes out of a preoccupation with waste. Like if we want a bigger and better car we are going to have bigger and better waste production" (1996, 303).

Used items have always been at the center of social economic transaction places. People found in association with waste and used commodities are social rejects—homeless, poor, and ragpickers. These people may not have chosen to live off and reside near used resources, but there are others who rejoice in waste places. Unofficial places where waste accumulates—flea markets, junkyards, and dumps—retain a peculiar mix of disgust and fascination. Lynch considers these places ripe for play, action, and fantasy. They evoke varied sensations and contain enduring delights. Children, less inhibited by accepted ideas of beauty, value, and cleanliness, like junk. They find much to explore in it, seeing it as diverse and stimulating (Lynch 1990, 153). In fact, during the 1950s, a new type of playground was built: the adventure playground modeled after a junkyard. Although it proved popular with children, for adults it raised questions of management, liability, and aesthetics. In the United States, it has become almost extinct.

Failure to notice waste, misconceptions about waste, and repulsion toward waste prevent us from deciding how to manage it well. They hinder our ability to make waste a meaningful part of our lives and to shape culturally significant waste places.

Dirty matters

Waste has always been encumbered by the concept of dirt. The two, waste and dirt, are undoubtedly interrelated. Waste is always dirty; dirt is (most often) waste. "Dirt" is often used as a synonym of "waste" in casual language. Like waste, dirt is elusive, ambiguous, and dialectical. There is no such material as absolute dirt. It is relative and can be many things, as we choose. Moreover, cleaning dirt and waste in one place only breeds pollution elsewhere. In *Plumbing* (1997) Nadir Lahiji and D. S. Friedman posit: "Behind every cleanliness resides an execrable uncleanliness—clean dissimulates unclean" (1997, 36). Thus, "the cleaner you are, the dirtier you are."

Dirt originally referred to excrement and to a mix of wet soil and filth. Its scope then expanded to include any foul, putrid, worthless, or filthy substance (*OED* 1989). But dirt is mostly defined in relation to its opposite, cleanliness. As Lahiji and Friedman point out, "Clean and unclean do not exist in real opposition as two positive facts. Rather they are two poles in a relationship of logical contradiction. The unclean is not a positive entity; it is only the lack, the absence, of clean" (1997, 36).

Both waste and dirt involve the feeling of pollution, a sense that something foul has attached itself to our body, things, or place—a feeling that is hard to shake. It is the fear of contamination, physical, medical, and moral. It is a fear of the traces of another human, an animal, or someone else's God, of "a kind of magical power that these traces might exert on us if we happened to touch them or even smell them" as Terence McLaughlin explains in *Dirt* (1988, 5). According to McLaughlin, we are tolerant of our own dirt and waste but fear that of others, since others might infect us with their illnesses and imperfections. Another scholar on dirty matters, Christian Enzensberger, makes the same point in *Smut: An Anatomy of Dirt,* claiming that dirt is fine between it and a person, but when a third party is involved, it becomes something unmentionable. "[It] threatens the proper separateness of the individual, his anxiously guarded isolation" (1972, 22). He posits that to grasp the idea of waste and dirt takes a fully developed sense of self; because children have no such concepts (or self), they will try everything imaginable.

The relationship between filth and self is masterfully expounded in another theoretical work, *Powers of Horror: An Essay on Abjection,* by psychoanalyst Julia Kristeva. Kristeva frames waste and dirt as abject and claims that the urge to reject the abject is fundamental to the construction of self and the "I." According to her the "ab-ject" is not an "ob-ject" outside the self nor is it an otherness that attempts to escape. "The abject has only one quality of the object—that of being opposed to *I*" (1982, 1). The abject is perverse and obeys no laws or limits. It is animalistic. In order to control it we must sublimate it and in the process develop a sense of borders, which define and separate the self from others (and from animals). A coherent ego, argues Kristeva, can only be established on the foundations of ejection and rejections of some of our own abject parts (feces, urine, sweat, semen, vomit, saliva, breast milk, etc.). We fear the abject because we know it cannot be completely eliminated; it looms from the border and threatens to return. Our borders are frail, and individuality remains ever unstable. "Abjection is above all ambiguity," adds Kristeva (9).

Abjection also has a relation to gender. It is predominantly associated with women, considered female property and responsibility—embodied in the female bodily production of abject substances and the role of cleaning and tending the household. But humans are all the procreators of dirt. In fact, they are made of dirt in and of themselves. As the Bible says, "You are dust [dirt], and to dust you shall return" (Genesis 3: 19). Hence, Enzensberger (1972) concludes, everything in the world is dirt; we are all dirt and all we touch is dirt.

Finally, dirt, like waste, is an outcome of a system of cleaning and or- dering (Douglas 1966). But while waste can be easily removed, dirt is hard to remove. The means of ridding oneself of or purifying oneself from dirt are burying or burning: "In the first case everything disappears by being immersed in an even greater dirt[,] . . . the dirt itself preserved; in the second case by disembodiment" (Enzensberger 1972, 15).

Human waste (i.e., feces), the most persistent and potent waste, is the most repulsive of all, and it is also the most ambiguous and dialectical. Excrement is equally bad and good, sickness and remedy, contaminat- ing and purifying, smelly and aromatic, and even ugly and beautiful. Hu- man waste is shit. And it is gold.

The peculiarities of shit

Human waste vividly conjures up our daily bowel movement, intimate moments of personal indulgence and relief accompanied by a private, delicate, and maybe even enjoyable smell. Human excrement is only one of twenty-five human bodily excretions, of which Enzensberger writes: "each one is a source of curiosity and pleasure. They are frequently pre- ceded by a feeling of tense expectancy and followed inevitably by con- tentment. . . . The fact is, man enjoys excreting. Then he rejects his handiwork" (1972, 7).[3]

Depictions of bodily fluids and excrement found in ancient illustra- tions present an ambivalent cultural view that projects not merely dis- gust and offense but also respect and reverence toward bodily cycles and their connections to the nurturing earth and to the gods. The Chinese, Japanese, Flemish, Romans, and, on and off, various Western peoples collected, composted, and used human manure. Many cultures dedicated gods to this important matter. Romans had the god of ordure, Stercutius (a name given to Saturn after he laid dung to fertilize the earth), and Crepitus, the god of convenience; Mexicans and Moabites had their own deities to honor the blessings of dung's fertility.

The Russian literary theorist Mikhail Bakhtin's influential work *Rabelais and His World* (1984) dwells on the ambivalence of excrement, claiming that in folk legends feces were closely related to fertility and birth. Rabelais's language often combined God and excrement. He attributed the birth of the springs of Italy and France to human urine and the birth of the Orion (in Greek, to urinate) to the urine of Jupiter, Neptune, and Mercury (Bakhtin 1994, 211). But the same anal functions are also the source of shame and contempt. Bakhtin's writings centered on the ambiguity of the lower body, its debasing and regenerative connotations and functions alike: "To degrade also means to concern oneself with the lower stratum of the body, the life of the belly and the reproductive organs; it therefore relates to acts of defecation and copulation, conception, pregnancy, and birth. Degradation digs a bodily grave for a new birth; it has not only a destructive, negative aspect, but also a regenerating one" (206).

In Bakhtin's medieval carnival human excrement plays a major role, engendering laughter about life and scorn for the ruling power. "What makes shit such a universal joke is that it's an unmistakable reminder of our duality, of our soiled nature and of our will to glory" maintains cultural and art critic John Berger (1991, 40). Post-Renaissance Western art has generally treated the human and animal waste within the context of, the comic and the grotesque, "the power of excrement to arouse laughter and its capacity to shock, repulse, and alienate" (Chu 1993, 41). In the nineteenth century artists and caricaturists, such as Emile Zola, employed scatology to criticize those in power and condemn their misdeed.

Human waste has an abundance of synonyms. Manure, sludge, dung, excrement, ordure, feces, and the medically favored stool and urine are but a few. Slang, which helps complete the picture, includes crap, piss, pee, pee-pee, ca-ca, and shit. Of all these, "shit" is the most forthright term (and the word choice of Laporte). So it remains that "In all languages 'shit!' is a swear word of exasperation" (Berger 1991, 38). Although mostly negative, some terms carry a positive tone. The archaic

use of the verb "to dung" meant "to cultivate," and even in contemporary definitions, "human dung" refers quite positively to fertilizer (*OED*
1989).

Laporte shows that the range of the history and politics of human
waste far exceeds our wildest imaginings. According to him, "Until the
very eve of clinical medicine, it was maintained that shit had the potential to be unquestionably good" (2000, 36). Early medicine and related fields emphasized shit's therapeutic and cosmetic properties. Texts
of antiquity claimed that shit cures wounds and disease. Accordingly, the
good and the beautiful were placed together with the excremental. Excrement was an equally effective remedy against the sterility of both
body and soil. Sewer workers, for example, sustained fewer rashes and
skin diseases than others, and their cuts healed quickly. Laporte maintains that the nineteenth century rediscovered the use of excreta as a
principle of life—a remedy and even a balm, a lotion of beauty's bath.
He argues,

> Civilization does not distance itself unequivocally from waste but betrays
> its fundamental ambivalence in act after act (32). . . . To this very day,
> civilization's ambivalence toward shit continues to be marked, on the
> one hand, by a will to wash those places where garbage collects (i.e., in
> city and speech) and, on the other, by the belief in the purifying value
> of waste—so long as it is human (37). . . . That which occupies the site
> of disgust at one moment in history is not necessarily disgusting at the
> preceding moment or the subsequent one (32).

Nowadays, human waste is a suppressed subject matter, long expelled
from both public space and language. It is an object of utter disgust. Despite and probably because of this, it has been equated with gold. This
notion began long ago, when humans discovered excreta's fertilizing
property, and it continues in the recent estimate by the Environmental
Protection Agency (EPA) that the sludge generated annually in the
United States is worth $1 billion. In his theory of anal eroticism, Freud
elucidates the connections between excrement, gold, and pleasure,
claiming, "It is possible that the contrast between the most precious substance known to men and the most worthless, which they reject as waste
matter ('refuse'), has led to this specific identification of gold with fae-
ces" (1977, 214). Freud argues that the ultimate civilized desire for
money replaces the repressed anal eroticism of childhood, which is tied
to unavoidable and in fact pleasurable bodily functions. These reserved

realms guarded by strict social taboos beg for a rescue. Laporte refers to the golden metaphor of shit as well. He recounts the many ways in which shit has been recycled into gold. In antiquity the Roman emperor Vespasian imposed a tax on urine. He placed vessels for relief at street corners and rented them to proprietors. Because shit has been aligned with the individual and defined within the private sphere (a process that began in the sixteenth century, claims Laporte), shit could enter the public only as gold, as odorless utility:

> Thus, as a "private" thing—each subject's business, each proprietor's responsibility—shit becomes a political object through its constitution as the dialectical other of the 'public.' . . . It is the home of that small heap of shit, which the subject tends to, maintains, even cherishes. The State, on the other hand, is the Grand Collector, the tax guzzler, the *cloaca maxima* that reigns over all that shit, channeling and purifying it, delegating a special corporation to collect it, hiding its place of business from sight (Laporte 2000, 46).

Cesspool collectors and many loyal believers in the medicinal properties of shit staged angry demonstrations against the grand nineteenth-century Parisian sewers. But the forces of modernism left no choice, subduing any of the subject's prerogative over waste.

Beyond shit's undignified psychological, visual, and symbolic attributes, it is the smell that really matters.

Waste smells

A loaded wagon of human manure (or its contemporary equivalent, the garbage truck) is a convincing argument for claiming one's right-of-way. Probing human responses to various types of shit, Berger notes that people do not resent cow and horse dung as much as human, and in fact can become nostalgic about them. Cow shit is even considered sacred by some cultures. In contrast, chicken shit, saturated with ammonia, is unbearable. Pig and human excrement, however, smell the worst because men and pigs are carnivorous and their appetites are indiscriminate. It is a sweet decaying smell (Berger 1991, 38).

The agent of entropy—the bacteria that labor to break down the sludge—is also the agent of the foul and putrefying gas. We tend to consider smells as an offense to personal sensibilities. But these feelings are conditioned by irrational fears. The smell of decomposition of anything we call waste is quite threatening. Generally, it is agreed, such smells not only represent a danger to health but also signify the degradation of matter, and that decline of value entails a social stigma. Smell is difficult to control. It penetrates our private space and transgresses social boundaries (fig. 1.5). Hence, as with cleanliness and order, the social regulation of smell is tight. It is systematically suppressed by culture because, among other things, it is a threat to social order.

During the past few centuries the development of the sense of smell—the sense of self-preservation—has been inversely related to the development of sight—the sense of the intellect. Sight became associated with reason; smell, with savagery and madness, animality and sensuality, lust, desire, and impulsiveness. Responding to the affront on smells of all kinds, Renaissance philosopher Michel de Montaigne argues in his 1570s essay *On Smells* their value for self-indulgence, culinary joy, medicine, and spiritual contemplation (1948, 228).

Yet, the passage into the Enlightenment could not occur without a refinement of the sense of smell, which entailed lowering the threshold of tolerance for certain odors. The German philosopher Immanuel Kant considered smell the most dispensable and the lowest and most subjective of all senses; it was invasive—like taste it penetrates the body, but we cannot control what we smell—and related to self-indulgence rather than to knowledge. Vision was a necessary source of truth and distance (Klein 1995, 20). Therefore, society invented ever-greater taboos on smells and increased the desire for odorless environments. In *Aroma*, Constance Classen, David Howes, and Anthony Synnott explain that "smell has been marginalized because it is felt to threaten the abstract and impersonal regime of modernity by virtue of its radical interiority, its boundary-transgressing propensities and its emotional potency." The authors claim that "in the premodern West, odors were thought of as intrinsic 'essences,' revelatory of inner truth. Through smell, therefore, one interacted with *interiors*, rather than with surfaces, as one did through sight." Such a sensory model was seen as an obstacle to our "modern, linear worldview, with its emphasis on privacy, discrete divisions, and superficial interactions" (1994, 5).

Medicine and science became seriously concerned with smells from the eighteenth century onward. The miasma theory focused on "bad air." Cesspool clearing made city streets stink and was frequently criticized. In *The Foul and the Fragrant*, Alain Corbin recounts the growing public debate on odor and the French obsession with perfume and aromatics. People carried a smell box in their pocket. Strong smells of musk, ambergris, and civet were used to "correct" the air and personal exhalations. Later on, the bourgeois preference shifted to natural floral scents, leaving animal scents to the masses. The elites were able to escape the bad smells of the city by traveling to the antitheses of putrid places—mountains and gardens, which were thought to have remedial properties (1986, 72). Fearing the emanations from the ground, cities searched for means to seal off the filth and muddy soil. Hence, in addition to fumigation and the use of salt and sulfuric acid to rid of smell, late-eighteenth-century urban projects began draining the streets via open and underground drains, as well as paving and cleaning them. These acts along with ventilation—wider streets and green, open spaces—became the main tactics to get rid of smells. Land use segregation and moving the offensive sources outside were others (Corbin 1986, 101).

Still, by the late nineteenth century humans associated smell with terror. In 1864 the renowned British sanitarian Edwin Chadwick observed, "All smell is disease" and "All smell of decomposing matter may be said to indicate loss of money" (quoted in Reid 1991, 56). Accordingly, hygiene and pleasing fragrance reassure us of health and buttress the sense of stability, and it is commonly agreed that the beautiful and the rich must not smell. The sweet smell of success and money is no smell at all (Laporte 2000, 85).

Twentieth-century technology and hygienic standards have taken the threshold of smell tolerance to a new extreme. Throughout the Western world, odorless environments have become preferable and the deodorization of the body is a thriving industry. The ultimate social regulation of smells is manifested in the current battle of the anti-scent movement. It calls for banning all scents or perfumes and demands a chemical-free environment (in part responding to new allergies that people have developed) (Klein 1995).

Strong smells have been associated with animality, decay, and death, concepts deeply embedded in the ideology and taboos of waste and dirt.

The four taboos of waste and dirt

In her groundbreaking book *Purity and Danger,* the anthropologist Mary Douglas maintains that "our idea of dirt is compounded of two things, care for hygiene and respect for conventions" (1966, 7). Whereas hygiene is directed at avoiding epidemics and death, conventions concern complex cultural issues. Although the rules of hygiene change as our knowledge grows and conventions change according to shifts in practical need and technology, there are some enduring ideas and taboos at the core of our relationship to dirt or waste. According to Douglas a reflection on dirt involves reflection on the relation of order to disorder, purity to impurity, form to formlessness, life to death. These four elements—disorder, impurity, formlessness, and death—represent dangers, are defined as taboos, and are averted by rules and rites of cleanliness. The four taboos are distinct but often confused: disorder is concerned with social structure and control; impurity, with morality and holiness (unity of experience); formlessness, with clarity of pattern and aesthetics; and death, with decay and finite being, one of the deepest human fears.

Disorder. "Dirt is essentially disorder," contends Douglas (1966, 2). Waste and dirt are matters out of place, "For what is misplaced in the house, and breeds pestilence, is well-placed in the field" (Reynolds 1943, 267). From another viewpoint, waste is a by-product of the creation of order; the unwanted bits of whatever it came from—body, food, wrapping. In both cases, waste and dirt constitute disorder, an offense against order (Douglas 1966, 2, 162). Every culture uses the idea of waste or dirt to order an apparently chaotic world in a distinct way, to define boundaries, and to achieve social and physical stability (160). This concern with divisions and boundaries—conceptual, physical, and social—is coupled with an attempt to embed, manifest, and perpetuate the hierarchy of the established and desired social value systems within strict spatial bounds.

As a civilized culture we expect to seek order and cleanliness. As Freud observes, "Dirtiness of any kind seems to us incompatible with civilization. The same is true of order. It, like cleanliness, applies solely to the works of man. But whereas cleanliness is not to be expected in nature, order, on the contrary, has been imitated from her" (1961, 46). Douglas shares this observation and points out that threats are pushed outside: "Pollution behaviour is the reaction which condemns any object or idea likely to confuse or contradict cherished classifications" (1966, 37). But Douglas also emphasizes the positive side of cleaning and ordering, "In chasing dirt, in papering, decorating, tidying we are not governed by anxiety to escape disease, but are positively re-ordering our environment, making it conform to an idea" (2). Finally, she suggests that as we seek order we also appreciate disorder. Disorder signifies danger and power alike; it has the potential for indefinite new patterns and order.

Impurity. Perceptions of waste or dirt have much to do with purity, be it material or symbolic. What is impure, how it can be avoided, and what rites are appropriate for cleansing are major preoccupations of holy teachings (Douglas 1966; Lynch 1990, 12). Religious doctrines equated waste with impurity and wrongdoing. Secular filth become defilement as it approaches the sacred. Holiness and impurity, therefore, cannot co-exist (Douglas 1966). The New England Puritans were among those who associated filth with sin. Sickness, especially related to skin and bodily deformation, was considered evil. For centuries societies rejected people with leprosy, believing they were possessed by an evil spirit. Rituals of cleaning were mostly enlisted to keep things pure and proper. The many Jewish cleaning laws conflate hygiene with obedience to God. But the rules regarding food and food containers in orthodox Judaism do more than serve a hygiene purpose: they help forge social identity, distinguishing the chosen people from the rest. Indeed, religion not only warns us about what is biologically unsafe; it instructs us to place boundaries between us and them, the moral and immoral worlds, good and bad, right and wrong. Judgments of purity and impurity set those boundaries and make the divisions distinct. Rituals of purification are also a guard against change or disruptions of moral codes. "Purity is the enemy of change, of ambiguity and compromise" writes Douglas. "Most of us indeed would feel safer if our experience could be hard-set and fixed in form" (163). And although "we are pattern and distinction-making creatures," according to Lynch, "we are not committed to maintaining fixed ideas of purity and value" (1990, 40).

Formlessness. The act of cleaning also involves an act of form giving. As Douglas suggests, "it is a creative movement, an attempt to relate form to function, to make unity of experience" (1966, 2). Aesthetics is based on the idea of the material world as a series of perceptible forms and patterns that contrast with their chaotic, formless surroundings. Waste lacks pattern; in fact, it spoils pattern. Some say that it is an attack against form (McLaughlin 1988). Form and pattern imply limit; from all possible forms and from all possible relations a limited set has been selected (Douglas 1966, 95). Thus, formlessness by implication is unlimited; like waste, it disrupts accepted formal arrangements and offends our aesthetic sensibilities. Formlessness also constitutes an affront on images that conjure up familiar meanings. The danger of pollution-creating behavior lies in its attack on familiar patterns and meaningful experiences. We accept and comprehend familiar patterns; we are cautious with ambiguous patterns and tend to relate or harmonize them with those we know; we reject discordant patterns (Douglas 1966, 37). But we do not always find formal ambiguity unpleasant. We enjoy works of art because they allow us to perceive forms that go beyond the familiar and the articulated. This observation brings us close to the final point Douglas makes, "There is power in forms and other power in the inarticulate area, margins, confused lines, and beyond the external boundaries" (99). For Douglas dirt and waste are symbols of creative formlessness; "formlessness is therefore an apt symbol of beginning and of growth as it is of decay" (162).

Death. Dirt and waste are evidences of the imperfection of life, a constant reminder of entropy and decomposition. Waste and dirt imply an unhealthy, out-of-control, unstable condition, a state of decline and degradation, and ultimately death. Most religions established rites and concepts to soften the inevitable. They also use the concept of afterlife to guard against deviance and uncleanliness through reward and punishment in the life after death (Douglas 1966). An absence of visible waste gives the illusion of a perpetual, unchanged world, though in reality such a condition denies the passing of life and the succession of generations.[4] The fourth taboo aims at helping defy the natural processes and functions of decay and death. It reflects a desire to counter entropy, to bring natural processes to a halt and achieve permanence and stability through intensive maintenance and cleaning.

The impulse to halt aging and prolong life is a fundamental and universal principle of human psychology. Of the four taboos, it is the most difficult to attain and utterly threatening.

The dialectics of entropy

The phenomenon of entropy, the supreme law of nature that governs the flow of energy in the world, is described in a variety of ways. In common language we refer to "decay" and "decomposition," the objective, physical terms that express the tangible disintegration of organic and inorganic matter, and to "decline" and "obsolescence," the subjective, cultural terms that convey loss of value, often as gauged by the marketplace. In thermodynamics, where the term originated, entropy (the second law of thermodynamics) is characterized in terms of the tendency of molecules of gas to spread from one corner of a container to its entirety, or the tendency of hot objects to cool down and match the temperature of their surroundings.[5] The law says, in essence, that everything in the entire universe began with structure and value and is irrevocably moving in the direction of random chaos and waste (despite the first law of thermodynamics, which maintains that matter and energy remain constant and are never lost). Entropy implies irreversibility; matter and energy by nature change only in one direction, from usable to unusable, from available to unavailable, from ordered to disordered. Statistical theory and the social sciences have also adopted entropy as a key concept. In statistical terms, entropy equates with disorder and denotes the improbability of increased organization in the world. Social entropy is used to measure the natural decay of social structures or the disappearance of distinctions within the social systems when their structure is not maintained.

Entropy, however, has been given ample and at times contradictory interpretations in an attempt to explain the increased sophistication and balance found in organisms and natural and cultural systems. Information theory, for instance, invented the term "negentropy," which means information that contributes to a coherent structure or meaning. It suggests that the world is moving to states of increasing order and information on a macro-scale. Negentropy measures the complexity of physical structures (e.g., buildings, organisms) in which energy is invested and which can become more complex by feeding on negentropy (for example, the natural cycle of decomposition, regeneration, and succession) (Schrödinger 1967). The dialectics inherent in nature and culture alike imply that everything, from basic cell to the organization of society, operates according to both processes of negentropy and entropy. The search for organizing principles or patterns is ongoing in areas such as biology, neurology, and artificial life and intelligence. For lay people, however, the entropic condition contradicts the sense of survival and continuity and the sustenance of material value. For some, entropy destroys the notion of history as progress and negates the belief that science and technology create a more ordered world. The primal instinct, therefore, is to oppose it, to exercise our power to ease and slow down the process of degradation in order to uphold the values, productivity, and usability of things and places.

In the past three decades the concept of entropy has become central to environmental concerns. Jeremy Rifkin's best-seller *Entropy: A New World View* (1980) connected the law of entropy with society's worldview and spiritual way of life. Rifkin claims that our lifestyle enslaves us to greater consumption and technological dependency. Warning of the planet's resource limits, of humans' wasteful habits, and of consumptive technologies, he maintains that entropy is a condition we must accept and integrate into our way of living. Yet he prescribes a "low entropy" formula to slow down its working. This entails a frugal lifestyle that favors such practices as community and urban gardening, farmers' markets, wood-burning stoves, bicycles, eco-minded architectural firms, and alternative technologies of solar and wind power. These changes, according to Rifkin, will bring independence and ensure spiritual freedom.

Others approached the subject of environmental degradation with a less regressive and moralistic tone. For them, dependence on advanced technology and current rates of consumption are not necessarily problems; the problem is lack of creative physical and industrial design. Kevin Lynch, for instance, advocates wise planning and design practices that support change and smart waste management. He calls for improved production, for adaptive reuse, and for the design of buildings and landscapes with change in mind. Lynch's approach, both practical and visionary, is grounded in the inevitability of change and the contradictory human impulses toward preservation and destruction. As the editor Michael Southworth puts it in his introductory remarks: "Decay, decline and wasting are intimately linked, they are necessary part of life and growth; we must learn to value them and do them well" (Lynch 1990, vii). According to Lynch, to see life whole, we must attend to loss. We need "to learn to see continuities in the flux, the trajectories and the unfolding" (201). Lynch also calls attention to the aesthetics and delight of entropy: "Why not take pleasure in breaking things when they must be broken, make cleaning a joy, find compensations in decline, deal openly with loss and abandonment, see death as part of life?" (166).

Similarly, the renowned artist Robert Smithson engaged in a passionate inquiry into physical decay and the philosophy and aesthetics of entropy. His entropy was the first of "many names for history's other," contends scholar Gary Shapiro (1988, 100). In a 1971 interview, Smithson (1996) told Gregoire Muller, "I'm interested in collaborating with entropy. Some day I would like to compile all the different entropies. All the classifications would lose their grids. Lévi-Strauss has a good insight; he suggested we change the study of anthropology into 'entropology.' It would be a study that devotes itself to the process of disintegration in highly developed structures. After all a wreckage is often more interesting than structure" (256).

Smithson used his art and writings to reveal and take part in the entropic experience. He opted for the fractures, sedimentation, and fissures of geology and archaeology, rather than the apparent continuities of the Hegelian history of spirit. "There is nothing more tentative than an established order," he claimed (1979, 107). Because entropy resides and conspicuously displays itself in the most marginal, isolated, ostracized, and maligned circumstances, Smithson sought out degraded, marginal landscapes—slag heaps and strip mines—as material for aesthetic exploration (fig. 1.6). He contrasted these infernal regions with the highly ordered ideal gardens of the past and their modern equivalents, national and large urban parks, which display a finished, static response to nature. In his pioneering works *Broken Circle* (1971) and *Spiral Hill* (1971) at an abandoned quarry in the Netherlands, marginal landscapes remain ambiguous and open-ended. His *Partly Buried Woodshed* (1970), a sculpture at Kent State University, and *The Bingham Copper Mining Pit* (1973) reclamation proposal in Utah are interventions that do not attempt to arrest or reverse time but rather make it conspicuous. They replace "the fiction of permanence with the fiction of change" (Horning 1987, 77). But according to Shapiro, even Smithson's chaotic aspirations were coupled with the tradition of proper use and preservation. For example, Smithson himself wrote specifications for the maintenance of his *Partially Buried Woodshed* and wished it to last as a permanent artwork. Similarly, and more surprisingly, he wanted to elevate his submerged *Spiral Jetty* (1972) after it was inundated by the rising Salt Lake waters in Utah (Shapiro 1988, 100).

Both Lynch and Smithson, for different reasons, searched for the dialectics of entropic change. They sought to make it a visible element in a landscape where control is limited and resource is finite. They called for the aesthetics of both chance and control, the dialectics of entropy and negentropy to guide art and environmental design. Fascinated by derelict landscapes, in which waste converges and fully collaborates with the elements of decomposition and regeneration, they turned to waste sites and found them in geographical (and social) margins.

Marginal Landscapes / Waste Landscapes

Waste and margin

Waste and margin are inextricably related. The two are alike in many ways and are mutually dependent. Waste is always marginal, and margins almost always include waste; in fact, they invite waste. Waste is often put in marginal areas to await further transaction. As noted earlier, Mary Douglas posits that purifying activities or cleansing rituals delineate boundaries and create margins, both physical and symbolic. The cultural historian Susan Strasser argues that "Many examples demonstrate the importance of physical margins to a history of trashmaking and disposal" (1999, 7). Disposability, according to Strasser, is a process of displacement, a removal of matter to a slightly "better" place, to boundaries, marginal spaces. The act of disposal is an act of sorting and placing rejected material into rejected marginal spaces.

As with waste, marginality is a condition conferred and maintained by society—usually by the powerful "center," the established sociopolitical group. The margin, whether conceptual, social, or physical, connotes a space or condition close to the edge, limit, or boundary. "Marginal land" is defined in simple economic terms as land barely worthy developing, akin to wasteland (*OED* 1989). Both waste and margin remain outside what is considered necessary or useful, and they are perceived as inferior. Additionally, both waste and margin imply the feeling of pollution. Ideas of pollution were enlisted to bind central and marginal places to their allotted roles. Clean places have come to signify that which is idealized, pure, and established; dirty places connote waste, the physically unclean and formless, the morally impure, the ugly, uncontrollable, and degraded.

The margin, like waste, is inherently dialectical. It is useless and useful, negligible and significant, repulsive and attractive. These dialectics are clearly apparent in the word's varied dictionary definitions. Although on the surface marginality implies a condition of lesser or low value—subordinate, powerless, and less desired—a close look reveals that it is a likely source of newness, choice, risk taking, and critical perspective. In psychology, marginal pertains to the important edge of the field of consciousness. In monetary terms, a margin may be considered essentially unnecessary at the moment but useful as *a reserve for the future,* a provision for unpredictable contingencies or a sum used to cover the risk of a loss on a transaction or account (*OED* 1989). Even in its original and literal (as well as literary) notion as the margin of a page—the space between the extreme edge and the main body of written or printed matter that is often taken up with notes, references, illuminations, and the like—the margin holds an important, dialectical role. It becomes a space that serves to reaffirm the authority of the central text (or of any center, for that matter) while commenting on, elaborating, and ultimately changing it. Apparently, this last, most literal definition is also the one that most clearly speaks of the critical role of the margin as a guide to the center. According to the textual scholar Laura Kendrick, the margin of the page in old scripture and classic text provided a space for commentary "whose peripheral position declared its own difference and supplementarity while affirming the power and originality of the text occupying the central position" (1992, 842).[6] Ultimately, the margin altered the original text.

The concept of marginality, whether in relation to ideas, activities, artifacts, people, or spaces, commonly is understood in relation to the center. Traditionally applied to the sociopolitical arena, marginality "has been increasingly used in sociological parlance as synonyms of weakness, poverty, and ignorance to indicate—in a meaning taken over from economics—a position of deprivation (alienation) with respect to power, wealth, and culture," explains economist Sidney Pollard (1997, 9).[7] In many instances marginal groups or individuals were aligned with the "minority" or the "other." They inhabited the margin, the borderland of a relatively stable center of some sort. Probing social marginality, postmodern scholars and thinkers exposed and deconstructed this problematic, hierarchical relationship. They contended that while in a position of relative exclusion, the margin could also be a position of power and critique. Subordinated, excluded, and subject to the central power, the marginal is not necessarily of minor importance or little effect.[8]

Marginalia and cultural discourse

Notions of the marginal in postmodern discourse were to a large extent grounded in psychology, sociology, and philosophy. Psychoanalytic theory suggested that the human psyche is not an integrated whole but rather fundamentally incoherent and fragmentary, or "decentered" (Harper 1994). Thus, Jacques Lacan declared that one's consciousness is not the center of one's being but is ex-centric to oneself. In a sense he implied that everyone becomes ex-centric to one's self and thus incorporates the margin (Lacan 1968). In sociology, the decentered self has taken the form of a marginal person to whom a greater cognitive and ethical freedom is attributed. Georg Simmel's marginal stranger is provided with the critical distance required for clear consciousness and prudent ethics. Simmel argues that such a person maintains contradictory attributes of distance and nearness, indifference and involvement, which make possible creativity, freedom, and objectivity (1950, 402–8).

Likewise, the *flâneur,* a key figure in Charles Baudelaire's earlier writings on the city and modern life, is a man endowed with consciousness, acting as a mirror to the growing, anonymous city crowd (Baudelaire 1965). Walter Benjamin later elaborated on and described the flâneur as a figure who chooses the margin and remains aloof from the crowd: "The *flâneur* still stood at the margin, of the great city as of the bourgeois class. Neither of them had yet overwhelmed him. In neither of them was he at home" (1973, 170). For Benjamin, the distant observer has the perspective needed for social introspection and commentary. Therefore, while margins have their limitations, they also have their advantage of vision, one that is more distant from the centers of power and conventions of selfhood.

Postmodernism, whose pluralism is its key "ism," should not be credited with recognizing the marginal. Modernism's positivist utopia acknowledged it first (Yúdice 1988). The utopic impulse to elevate marginal, nonconforming subjects and unify society directed modernism to engage marginal groups—minority, immigrants, the poor, whores, beggars. But in such cases the marginal was seen as a problem that needs a cure, explains the linguist scholar George Yúdice. For Yúdice, the center's seemingly conciliatory intentions carefully guided and controlled the marginal while moving it toward the mainstream, which ultimately co-opted it. Thus, marginality did not last in modernity, nor was it resolved (Yúdice 1988). Likewise, Richardson's idealized city of Hygeia had no place for marginal groups (or marginal places); instead, they were assimilated into the center. In contrast, postmodernity both elicited and retained distinctions. "The postmodern tactician," claims Yúdice, "often *uses* the 'marginal' to make a case for his or her own subversive potential" (1988, 214). He or she preserves the distance and the awesomeness of the unfamiliar. Moreover, the postmodernist asks the marginal to keep its particularity and nurture the very condition that enabled it to survive, challenge, and serve as an alternative to the center (Yúdice 1988). Postmodern culture and art scholar Russell Ferguson elaborates: "[The marginalized] don't allow themselves to be defined only in relation to something else. They stand their own ground, and speak from there without apology" (1990, 9). For Ferguson, postmodernism should not create a new center of authority of the marginalized or hammer out the binary oppositions between central and marginal, "but rather . . . open up spaces for new ways of thinking about the dynamics of cultural power" (9).

Postmodernity also recognized that the workings and relations between center and margin are in fact never fixed or stable but shift among loci of subjectivity. There can no longer be a single recognized center or margin, elicits geographer Rob Shields. "[The margins] expose the relativity of the entrenched, universalising values of the centre, and expose the relativism of cultural identities which imply their shadow figures of every characteristic they have denied, rendered 'anomalous,' or excluded" (1991, 277). This was the critique that disclosed the resemblance of margin and center, their inclusion in one another. Cultural critic Michel de Certeau goes even further, claiming that "Marginality is today no longer limited to minority groups, but is rather massive and pervasive." Though consisting of varying groups, the general consumer of the new socioeconomic structures "has now become a silent majority" (1984, xvii).

One subtext of this broader discourse was geographic marginality. The narrower discussion articulated the significant role of marginal space.

Marginal land / waste land

Early on, geographic marginality was attributed to physical remoteness or to forbidding and extreme topographic and climatic conditions, such as distant islands, marshes, mountains, forests, deserts, and very cold regions. It constituted the fringe of civilization and implied a low economic value for the dominant powers. Therefore, a land that for one reason or another was unsuitable for inhabitation or cultivation—that is, any region deficient in food production—was likely deemed marginal (Pollard 1997).

But the forces engendering physical marginality have changed with time. Sidney Pollard, who examined the links between geography and economy in marginal regions of Europe from the Middle Ages to the nineteenth century, posits that in the eighteenth century, spatial marginality was viewed in terms that were essentially political and social, despite containing geographic and economic elements. Marginality was sometimes blamed on lowly and passive populations. But Pollard repudiates the negative perception of marginal areas, especially the view that in these regions people submissively received and did not create or possess cultural values. He claims that the very qualities of poverty and hardship encouraged a pioneer role. Thus, marginal regions, such as British Lancashire and West Riding of Yorkshire, made a crucial contribution to the industrial revolution, taking the lead in technical and commercial innovations in eighteenth-century Europe (Pollard 1997).

In the nineteenth century regional classifications were bound to become more complex, the result of human handiwork. Industrialization and mining and their resultant waste and pollution rendered some regions unhealthy, depleted of resources and energies, and, at least temporarily, weak and depressed. In the twentieth century marginal/waste land has increasingly become a mental condition that implies desolation and alienation, which in turn led to abandonment and degradation. This phenomenon is best conveyed in T. S. Eliot's *The Waste Land,* published in 1922. Seen alternately as the founding text of modernism or postmodernism, the poem expresses a new and radical, even nihilistic, state of mind, and an awareness of a historical change embodied in a mental waste land. Colored in gray and brown, immersed in despair, degradation, and mutation, and silenced, Eliot's waste land represents a human experience of alienation:

> The river's tent is broken: the last finger of leaf
> Clutch and sink into the wet bank. The wind
> Crosses the brown land, unheard. The nymphs are departed. (1949, 33)

The lines magnify the contradiction between reason (the underlying tone of modernism), which desperately seeks order, and vehement lust, which interferes with that search. Unending displacement, confusion, and destruction represent the sterility of the modern world and hint at a desire that lies underneath it.

More recently, new wastelands have been created closer to home. Old ports, postindustrial lands, and downtowns that lost their value owing to economic shifts, harmful human practices, or contamination have become the new margins. Urban districts and even whole towns and regions were left behind in the modern race for progress and became margins. New suburbs were hastily appended to metropolitan peripheries, and sprawling edge cities accommodated the fleeing urbanites. Soon, these seemingly new centers likewise became marginal lands. American middle-class suburbia and edge cities have turned into desolate loci, contemporary mental wastelands, claims Lee Morrissey (2000). The contemporary discourse on margins and wastelands has thus expanded to include the old and new centers (or the new and old margins) simultaneously, the entire everyday landscape. The peripheral dimension and its concomitant waste landscape, as in Eliot's mental alienation, concern everybody.[9]

Though margins have always shifted and evolved, the presence and potency of some marginal conditions in the landscape is constant.

The role of marginal/'other' places

Marginal places have always played a key role in the matrix of the landscapes. Probing large-scale, marginal areas such as the British seaside resort of Brighton or the Canadian north, Rob Shields posits in *Places on the Margin* that despite their marginality, these places have continued to evoke both nostalgia and fascination. Located at the periphery of cultural systems of space, they constitute zones of Otherness and precariousness, sites of illicit social activities. They signify the polar opposite of great cultural centers and expose and complement places at that center, where only accepted norms and behavior are permitted. Shields argues that contemporary Western society has always been attracted to the otherness of marginal places in a search for unfulfilled desires in the established, constructed identity of the center. At Brighton modest manners and polite plays in decent Victorian halls were replaced with uncovered bodies and street theater. Shields associates Brighton's beach with Mikhail Bakhtin's idea of the medieval carnival—a liminal zone of spectacle, feat, and celebration characterized by vulgar laughter and language—where people participated equally in an inversion of the official high culture's rules and manners (Bakhtin 1984). For Shields, "'margins' have become signifiers of everything 'centres' deny or repress" (1991, 276). At the same time that the center rejects the debasing "low," it also desires it.

The power of Other places to repel, lure, and fascinate us, as well as their potent, subversive role relative to the center, was the topic of a 1967 lecture by Michel Foucault. Foucault discusses a group of sites he calls "Other Spaces," or "Heterotopias."[10] He divides all spaces into ordinary and extraordinary. The extraordinary are further divided into Utopias, or unreal places, and Heterotopias, which are real. Heterotopias, claims Foucault, have a special relationship to all the ordinary places that people create; they both expose and contradict ordinary places. Among other types, he speaks of Heterotopias of Deviation, where individuals whose behavior is deviant in relation to the required mean or social norm are contained, illustrating his point with examples such as a prison, a psychiatric hospital, a cemetery, and a brothel. According to Foucault we create these places to contain and control all social perversions and threats and dread them when they go out of control. Heterotopic places also include other types of institutions, such as museum, zoo, fair, and garden, places that are positioned at the other pole of the continuum. Other Spaces, contends Foucault, relate to all everyday places in a curious way. They contain the everyday and yet they subvert, contest, and often threaten those places, thus providing an excuse for their own removal and rejection. Most importantly, Other Spaces shed light on our everyday places. They both affirm and contradict the everyday or normal *and provide us with the distance needed to look critically at our common landscapes and habits.* He writes, "On the one hand they perform the task of creating a space of illusion that reveals how all of real space is more illusory, all the locations within which life is fragmented. On the other, they have the function of forming another space, another real space, as perfect, meticulous and well-arranged as ours is disordered, ill-conceived and in sketchy state" (1997, 356).

Waste sites—landfills and sewage plants—easily fit Foucault's category of Heterotopias of Deviation. Sequestered in remote, fringe locations, they too contain and treat undesirable material culture or conditions and are most successful in playing the social roles of landscape mirrors, supplements, and critics.

Waste landscapes: A design framework

The paradoxical nature of waste landscapes—the peculiar mix of disgust and fascination they arouse—has engaged many writers. In his early work the landscape scholar and journalist Grady Clay identified an anonymous type of place he found in every city; he named them "sinks." These landscapes conjure up physical and mental images of depression, are characterized by geographic awkwardness—topography, size, or form that make them uninhabitable or undesirable by established standards—and subsequently are turned into the receptacles of society's detritus (fig. 1.7). Clay defined them as "places of last resort into which powerful groups in society shunt, shove, dump, and pour whatever or whomever they do not like or cannot use: auto carcasses, garbage, trash, and minority groups" (1973, 143).

Tucked in corners and carrying a bad reputation, these fragments, slivers, leftovers, oddments, and low basins invite trash dumping and dirty transactions. Yet they are also sites of opportunity, where new energies lie dormant but potential and where inventions, innovations, risk-taking, and imagination can find an arena free of the bounds of tradition and moral prejudices (Clay 1973). In a later work, Clay discusses a group of places, including abandoned farms, depressed areas, dumps, and ruins, under the category of Out There. Remote from and disregarded by the center, breeding suspicion, stinking and stewing, and attracting social outcasts, Out There, he claims, is a powerful, alienating force in American politics, as well as the origin of countless booster enterprises. Yet, "Out There needs The Center's help and participation as badly as The Center needs Out There. Neither is self-sufficient" (1994, 176). Lynch also distills the power of waste landscapes and describes them as places that escape the weight of power, the intent to impress, the sense of immediate human purpose. They are liberated zones, released from control, displaying rich boundary conditions and highly expressive forms (Lynch 1990).

Conversely, environmental design and planning, the tools society uses to "treat" and shape its marginal and waste landscapes have until recently overlooked the opportunities. The theory and practice of modernist environmental planning and design reduced the potential of marginality, erasing marginal places' particularity and source of power. Marginal lands, such as dumps, slaughterhouses, and slums, were targeted for "recovery," whose tactics included desegregation, urban renewal, and the construction of social housing projects and genteel parks. Swamps and sinks were bulldozed and covered with freeways, and decaying urban neighborhoods razed and their inhabitants uprooted.

The two typical modern approaches for waste facilities were camouflage and utilitarian (Engler 1995). The *camouflage approach* paid tribute to the urban aesthetics established in the late nineteenth century and continued the tradition of disguising waste sites. It used screening or cloaking techniques to conceal waste facilities. Dumps were covered with bucolic parks, and sewage plants were surrounded with berms and screened with heavy plantings. The *utilitarian approach* reused the waste sites as public amenity for recreation, for agricultural, and sometimes for private land development. The waste, an integral part of the sites, did not inform the design. Instead, the waste lands were recycled into generic parks that concealed their former phase. Such approaches presupposed that marginal landscapes were negative and threatening, the opposite of pristine, cherished landscapes. They hoped to suppress differences and create social and landscape unity.

Postmodernism took the old myths of marginality and turned them on their heads, endowing them with a positive and subversive power.

But in so doing it often fell into the pitfall of a revived positivism, "of re-essentializing oppositions from the 'side' of the marginal, and of prematurely celebrating differences rather than continuing to deconstruct difference" (Crewe 1991, 123). Consequently, postmodernist environmental design theory—still the dominant mind-set today—followed two distinct and parallel tracks. Both valorized and empowered marginal places, but the first viewed the relation between center and margin in opposition while the second saw the relationship as dialectical.

The oppositional track acted on the pervasive social and ecological despair, viewing nature and culture, and natural and technological forces, as contradictory pairs. As it reclaimed the marginalized (from its inferiority) it hammered out a new set of oppositions. The green heart of environmental designers and planners led to an impressive list of projects: the surfacing of buried creeks, the reconstruction of lost wetlands, cleanups of toxic land and water, and the resurrection of nature reserves on dumps. In these projects nature (the oppressed) often remains in opposition to culture, city, and technology (the oppressor). The two typical design approaches that have informed this track were restoration and mitigation.

The *restoration approach* considered the disposal function as a temporary land use disturbance that needed to be fixed. It sought to rehabilitate damaged sites by returning them to their previous conditions. Restoration projects simulated a historical, selected landscape, sometimes at great expense. Restoration did create viable new landscapes for wildlife and reconstructed "nature" for people, but it was just a different form of masking the waste that people had a hard time to accept and manage.

The *mitigation approach* sought to weaken the impact or reduce the severity of a polluted land or water. Designs emanated from scientific solutions. Informed by the process at work, the approach was based on the understanding of how nature works, and it sometimes resulted in life-supporting landscapes that attracted people and other species. The landscape designs evolving from this scientific-engineered solution are truthful to their function, yet without verbal interpretation their content and process remain obscure and inconspicuous.

The dialectical track of postmodern environmental design—a minor but important voice—took marginalia and continued to guard and deconstruct its differences, to dwell on its various relationships with the center, and to preserve its peculiarities. Obsolete industrial zones were reopened as museums; industrial ruins were revived according to their own daring prescriptions and meanings, lending themselves to adventure and play; and toxic sites were planted with contaminant-absorbing plants. Landfills were rebuilt according to the specifics of the waste place and the technology used to shape them, rather than turned back into a nature reserve or a city park. Some were shaped as cultural monuments and ritual sites. Wastewater and recycling centers were designed as spectacles and accessible public spaces. The intent of these designs was to avoid the binary oppositions of marginalized and center. Their goal was to open up spaces for new ways of thinking about the dynamics of cultural power and habits, as well as about landscape value and aesthetics. These planning and design approaches aimed to become catalysts of change and to create a diversity of voices that help disrupt the unified, commanding voice of authority imposed by modernism, the voice that subsumes all differences under its leveling cloak.

The two characteristic design approaches of this dialectical track were educative and celebrative. The *educative approach* emphasized public awareness and worked to change attitudes toward waste. It invited people to experience the realities of waste institutions and nurtures a more open relationship toward refuse. The *celebrative approach* promoted and dramatized waste sites and facilities through works of art, special design features, and unique experiences. The garbage or its processing was highlighted and iconized, provoking thought and meditation. The garbage became a metaphor for excess and resource mismanagement. The celebrative approach revealed to the naked eye the ongoing processes taking place at waste treatment and disposal sites. For adherents to this dialectical track, marginal landscapes performed a particular political and aesthetic function in relation to central landscapes and served as alternative places of opportunity, preserving heterogeneity and fractures. They represented a clashing plurality of landscapes.

The emerging supermodernist environmental design agenda of the 1990s preserves the general intent of the postmodernists while endowing projects with greater efficacy. Taking advantage of sophisticated ecotechnology and responding to clues about new social patterns and environmental problems, supermodernism offers flexibility, adaptability, multilayering, and functionality. These designers divert attention from marginal landscapes, already valorized by postmodernism, and focus instead on everyday landscapes. Waste landscapes are decentralized, reconceived as normal, practical, publicly accessible, and productive landscapes that are integrated into the everyday vitality of public life. Supermodernism represents a resilient view of reality, which seeks workable schemes in which people play a creative and participatory role. Heavily polluted areas are set off-limits or are seeded with hospitable neophyte plant species, left to grow with only casual monitoring. Dump reclamations are devoid of romantic urbanistic and technological formulas or references. They are used for a variety of experiments of extreme habitats or re-incarnated as waste processing centers. Reuse and recycling programs become community centers, new types of consumer-friendly meeting spaces, and loci for social action.

The supermodern brought forth two approaches: sustainable and integrative. The *sustainable approach* is concerned with the economics, conservation, and self-sufficiency of a waste disposal or processing site. It employs a diverse program that often includes elements of production or reuse of renewable waste resources. The sustainable strategy considers waste to be a valued resource. The waste operation is made somewhat conspicuous and mostly accessible to the public for activities that involve both waste disposal and recreation.

The *integrative approach* is a multifaceted one, combining elements from the celebrative strategy with other strategies. It integrates the principles of ecology with the philosophy of art—scientific rigor with expressive metaphors. It is layered with information and meanings that express the dynamic balance of nature and culture. The integrative strategy changes an abused site or waste facility while amplifying its reality. It always accommodates public access and often includes an active production component or working facility to serve community needs. Creating visible connections is integral to this strategy. Its focus is creating a healthy, integrated human-nature ecosystem, where work and leisure coexist, and in which all parts of the system are equally visible. The integrative strategy expresses fresh spatial conditions and aesthetic possibilities, which help make waste and its consequences accessible to public perception. It combines ecological, economic, and social agendas and merges the utilitarian with the experiential. It invites people to partake in highly sensory and transparent landscapes. Above all, it facilitates the flow of matter and the sustenance of vital waste landscapes.

Waste institutions have been lastingly charged symbols of social and economic neglect and failure; at the same time they contain the seeds for recovery. Re-forming these places has always represented both symbolic and practical acts with great repercussions. Although some environmental designers still cling to old ideas and formulas, a few begun experimenting with new formal, spatial, and programmatic possibilities that create more practical, lively, and congruous private and public landscapes.

CHAPTER 2

Private Landscapes of Waste

"Hygiene is the modern project's supreme act."

Lahiji and Friedman, 1997

Hygiene and Plumbing

Hygiene is the triumph of science and technology over the dark regimes of disease and filth.[1] But, as noted earlier, hygiene is not simply a set of practices and apparatuses to maintain health; it is a means to other social ends.

Society has considered hygiene an imperative for a moral and productive mind. In addition to its connection with ideas of health, order, purity, stability, and beauty, hygiene also has been linked to privacy, class, gentility, and economy. The history of personal and domestic cleanliness reflects a civilizing process. As the social historian Georges Vigarello observes in *Concepts of Cleanliness,* a study of the shifting ideas of hygiene in France since the Middle Ages, "[Hygiene] is a history of the refining of behaviour, and of the growth of private space and of self-discipline: the care of oneself for one's own sake, a labour ever more squeezed between the intimate and the social. On a wider plane, it is a history of the progressive pressure of civilisation on the world of direct sensations" (1988, 2).

Bathing is a case in point. For the most part, bathing in antiquity was not intended for cleaning—the management of bodily odor and the visible removal of dirt—but rather for religious or leisure purposes. Ancient and traditional baths, such as the Greek gymnasium bath, the Russian vapor bath, and the Islamic *hammam,* were all sites of regeneration rather than ablution, contends the cultural historian Siegfried Giedion (1948, 711). Ritual bathing was abolished as religion tightened its grip on and linked personal cleansing to ideas about water and the body. Through the Middle Ages, water was regarded as sacred and was not to be wasted on profane tasks; nudity was taboo. Bathing was also considered a threat to health. Water was seen as capable of penetrating the body; hot water was believed to weaken organs and leave skin pores open to insalubrious air. Soon after the Middle Ages, the concern for privacy, the social imperative to conceal nudity, and material demand for running water shaped the process of bodily cleansing. Bathing became a private, hygienic management of the body.

The private sphere begins with the body and extends to the house. Hygiene is the interaction of behavior with place and technology, and it is also the process by which particular things or places are deemed clean or dirty (Jenner 1997, 107). As Vigarello observes, "Care of the body involved a total reconstruction of the world above and below cities. . . . Cleanliness affected conceptions of towns, their technology, and their resistance to being criss-crossed by pipes" (Vigarello 1988, 230–31).

Plumbing required society to overcome fear of technology and build trust in modern mechanics. The task entailed major surgery in the bowels of the city and at a scale previously unknown. Before running water could be brought into the city and the house, pipes and a pressure system to convey the (clean and dirty) liquid had to be supplied. As cities throughout Europe and America installed water networks, more dwellings took advantage of sanitary fixtures.[2] These in turn became safer, less expensive, and more prevalent. The increased use of running water in homes caused cesspools to overflow and forced cities to proceed with building sewers (Tarr 1996). The 1870s were the climax of the American plumbing project. Most cities were busy laying out underground pipes. The sanitary engineering profession and trade magazines established their authority. The plumber ruled the day (Fig. 2.1). Plumbing embodied the idea of modernity—invisibility, central authority, efficiency, and hierarchy.

At the onset of the nineteenth century, hardly any American had or imagined a bathroom in the house. By the end of the nineteenth century about 50 percent of urban American households showcased a bath and a toilet connected to municipal water and sewer systems. In his 1898 essay titled "Plumbers," the Viennese architect and cultural critic Adolf Loos compares Austrian sanitary household fittings to American: "In this regard America is to Austria what Austria is to China. . . . A home without a bathroom! Impossible in America. The idea that at the end of the nineteenth century there is a country with a population of millions, all of whose inhabitants cannot bathe daily, would be outrageous in America" (1997, 17). He praises the English for their sanitary inventions, but goes on to point out American superiority. He equates civilized culture with hygiene and plumbing and postulates that the plumber is "the pioneer of cleanliness . . . the first artisan of the state, the billeting officer of culture, of today's prevailing culture" (18). Loos considered hygiene not only the secret of civility but also its source of power, and he claimed that the hygienic culture was bound to rule the world.

Yet the same plumbing that made modernity possible was also used to dismantle it. Plumbing became an avenue for the postmodern critique of modernity. In *Plumbing*, Nadir Lahiji and D. S. Friedman undertook "to plumb the depths of modernity" and exposed the limits of hygiene and plumbing. The limit was the vertical, hierarchical nature of the hygienic body (person or city) as opposed to a horizontal, interconnected system. Echoing some of Freud's early notions, they propose: "The modern discourse on identity, subject, space, gender, and body all presuppose the discourse of verticality, which reaches its limits in the repression of the common abject—excrement, putrefaction, dirt, semen, menses, and so on. . . . The superego of the hygienic movement constructs modernity by plumbing the destructive instinct of the pleasure we take in dirt and pollution" (1997, 8, 53). Ultimately, the vertical plumb line had to fail. Lahiji and Friedman countered Loos's plumber proposition with their own: "Plumbers travel between purity and abjection. They order everyday fluids, manage flow, straighten things out, keep things clean, sound depths, right columns, fix pipes: plumbing leads to the bottom of things" (11).

At the bottom of things there is waste. Technologies and rules of cleanliness were constantly used as social tactics, taming, containing, disguising, sanitizing, and finally redefining waste itself. As new weapons were devised, attitudes were changed, aesthetics modified, and landscapes reordered. Subsequently, the American residential landscape evolved to reflect and facilitate these changes.

Shifting Scenes in the Residential Domain

Early American householders relegated waste and cleaning functions to the outside of the house—yards, alleys, and streets. Since the 1700s, processes to contain, domesticate, and eliminate waste and margins have taken place. Places of waste and cleaning changed locations and nature; some were eradicated, others created. The presence of waste was instrumental in creating margins and in marking distinctions between indoors and outdoors, front and back yards, and street and alley until the waste was cleared away and transferred to the domestic sphere and then to the public underground. Waste was magically moved from the secretive places of the private to the hidden zones of the public. Consequently, the margins of the residential sphere shifted, narrowed, and almost disappeared altogether.[3]

In what follows, I sketch eight distinct periods of residential landscapes, each lasting thirty to fifty years. The primary consideration in determining the span of each period is the visible form and space of indoor and outdoor residential landscapes. The boundaries of the periods are not decisive, however, and some transitional overlap should be allowed. The first period, *accommodating the outdoors* (1770s–1800s), sees the entire area of the house lot and street used as a waste processing arena. *Cleaning the front* (1800s–1840s) corresponds with the removal of waste / cleaning functions to containment within the backyard in an attempt to project a clean appearance and a higher class status. *Containing waste* (1840s–1870s) coincides with the earliest mass plumbing and sanitary solutions and the desire of middle-class families for conveniences. *Diverting waste* (1870s–1890s) witnesses the ultimate domestication of waste functions and the transfer of waste to distant public locations in response to major scientific and technological developments linking waste with health.

In the *clean is beautiful* period (1890s–1910s), the front yard is transformed into an open, continuous, and tidy space and coincides with an increased tendency of the public to link cleanliness with beauty. *Sanitizing spaces* (1910s–1940s) describes a process of sterilization and the disguise of waste in the private domain, as affluence shapes wasteful consumption and leisure habits. *Out of sight, out of mind* (1940s–1970s) corresponds with the transformation of the yard into open, uniform, leisure grounds and the growing obliviousness of the public to increasing amount and threats of waste. *History and waste recycled* (1970s–1990s) traces yet another shift in the design of the private realm and the status of waste. As waste attained a new status, it became a resource and a symbol of a "green" ethic. Subsequently, the unit of house-yard reintroduced spaces for domestic waste processing. Current ideas, technologies, and practices regarding waste are promising still more changes. *The smart house and yard* (1990s–present) suggests that we are on a new threshold in the history of private waste landscapes.

Accommodating the outdoors (1770s–1800s):
Waste, a mere nuisance

Until the end of the eighteenth century the concept of waste, as we know it today, did not exist. All used or leftover matter was reused and privately managed. The traditional practice of reuse was a response to scarcity—the limited availability of energy and matter—and to the high costs of goods. It was intimately related to the idea of frugality and to conservation by the individual, who saved and repaired products and avoided discarding anything. The storing and processing of waste was done at the yard, where much of the family's food was produced (or in some basements). Without plumbing and running water, all functions associated with the removal, processing, and storage of waste, including bodily cleaning and human waste, were relegated to the outdoors—yard, alley, and nearby stream. The placement of waste, therefore, accentuated the distinction between indoor and outdoor spaces.

In the late eighteenth century the primary domestic generator of waste and the indoor place of "dirty" work, the kitchen, became a subject of concern and irritation. The early colonial kitchen (and its hearth, the only source of heat and food), which was the center of the house for cooking, eating, and other family activities, moved from its central status and location into the basement or into a separate wing in middle-class homes (Plante 1995). The messiness, odors, and servant work were pushed away from the social life and functions of the family. In the American South, the practice of removing activities related to food storage and processing away from the main dwelling into completely separate service outbuildings in order to eliminate the undesired kitchen heat, smell, and fire hazard continued well into the nineteenth century. Trash, such as broken ceramics, was buried in pits just beyond the outbuildings (Linebaugh 1994).[4]

Early sinks or washbasins were mostly installed in basements or "sinkrooms," next to the kitchen and the door to the back yard. A hand pump connected to a well or a cistern first fed tap water. For centuries, people had in bedrooms a simple freestanding washstand on which a metal or porcelain basin and water jug stood. In the late eighteenth century, bathing naked and alone was still considered frail and perverse. The smell of one's body was reassuring, and it was morally inconceivable to think of "getting wet all over at once" (Bushman and Bushman 1988, 1214). People used portable indoor tubs, wooden and metal, or containers filled with water for infrequent and partial bodily cleaning. As doctors began prescribing various forms of water cures for the ill, the vapor bath became popular.

American gentry, however, who had always been alert to cultural cues from London, began taking notice of their English counterparts' fascination with tubs and bathhouses served by water pipes. The first bathtubs and showers, connected to the improvised pipes and a boiler, were installed in "bathing rooms" built in the yard. In 1778 Henry Drinker, a Quaker merchant, installed a shower box for his family in the backyard of his Philadelphia town house, and five years later he added a tub. In 1796 St. George Tucker turned the dairy house at his Williamsburg residence into a bathhouse containing two copper tubs (Bushman and Bushman 1988, 1214). American tinsmiths began making baths with a variety of designs for indoor use, but only the elite could afford them.

For the removal of bodily waste, a privy-cesspool system was set in place. The privy was a small structure with one or more seats built over a cesspit where feces accumulated. American settlers called it an "outhouse" and took the facility with them as they moved westward. Privies were used extensively in rural America well into the twentieth century, and some can still be found today in the countryside. They acquired almost two hundred nicknames—one-holer, two-holer, convenience, necessary, johnnie, ajax, stool, pool, post office, potty, oklahoma, and house of parliament are but a few (Barlow 1989). Privies of the period were often connected by brick drains or earthen pipes to a cesspool or liquid manure tank made of bricks. Some were attached to a toolshed or a smokehouse. Wealthy residences could afford two compartments and multiple seating, as well as expensive interior wall finishing and an ornate exterior. Luxurious privies even boasted a fireplace. The privatization of waste made this small, discrete space a place for an inner monologue or even a confidential conversation with a friend. It might have been the only place one could escape family members and enjoy a quiet moment.

Ground holes or receptacles in outhouses or in dirt cellars contained the waste until they were cleaned or replaced. Cleaning cesspools was a private responsibility. In the country, when the ground around a cesspool became saturated, the privy was moved to another location, the cesspool covered over with dirt, and a new one dug; in the city, it needed regular emptying. The privy represented hardship—hard labor, filth, pollution, and, mostly, stench. To relieve the reek people used deodorizers with an equally strong odor of lilac, lime, lavender, or pine, and they planted lilies and lavender shrubs nearby. Animal manure and human waste was often sold to farmers and used as fertilizer. Otherwise, disposal of human waste was done carelessly into a dry well or a nearby waterway.

The early American yard, with its multiple work, storage, and processing functions, housed, in addition to the privy and vault-cesspool, a variety of structures and containers for water and waste: a well, a water pump, cisterns, and storage bins or platforms for wood and ash piles. The yard was a productive space containing a kitchen garden and domestic animals. Food scraps were fed to animals, composted, or used as fuel for heat. Storage spaces for the management of containers and bottles, rags and bones were set aside outside the house, often next to the barn, shed, or outhouse. Other spaces were allocated for items in need of repair. Only irreparably broken objects were ultimately buried in small refuse pits in the backyard. Compost bins were located in shaded margins. Furnace ashes were piled in yards and often applied to the garden or farmland to better the soil or used to cover human excreta in privies.

In the few cities near the turn of the nineteenth century, growth and congestion turned waste from a mere nuisance into a "problem." Households fronting streets treated them as front yards and disposed of their garbage as they saw fit (Schultz 1989, 115). Back alleys were frequently filled with ashes and garbage piles. Herds of scavengers—swine and chickens—commonly roamed the streets to feed on food remains. Private companies for garbage collection and cesspool cleaning were used only by the wealthy, and service was spotty at best.[5] Despite city ordinances, few people regularly cleaned and swept their frontage area. Cesspool contents and animal carcasses and manure in the streets became major annoyances—olfactory irritants that were hard to regulate and, more critically, a serious health hazard. Epidemics broke out often, but the disease was not attributed to waste matter itself. Rather, it was blamed on some chemical or physical influences transferred during direct contact with certain population groups and sick individuals.[6] The problem worsened, yet only a small number of cities enacted restrictions and regulations to control the timing and manner of disposing of and transporting the offensive matter.

The upper class brought running water and perfume to the rescue. The wealthy began installing improvised pipes carrying water from private sources to water fixtures located in outbuildings. The middle class, who could not yet afford this luxury, adopted new standards of space and habits of cleanliness to avoid the nuisance associated with waste and to project a genteel image to distinguish their status from that of the masses. Strong scents made from animals were applied to the body to counter and alleviate the pervasive stench. These and other deodorizing techniques gave waste and dirt a new power: to define and divide classes.

Cleaning the front (1800s–1840s): Waste and class

Waste and divisions of class became inseparable in early-nineteenth-century America. "Dirtiness," which always appeared in areas with the suspect qualities of impurity, immorality, backwardness, and ignorance, now justified social ranking of race and class. The clean appearance of people and their private domain—the visible facade—became a decisive criterion marking high status and gentility. The front yard was cleared of its waste and storage functions to make space for an ornamental garden, while waste was confined to the back. For indoor cleaning, householders were advised to use two water pails and a soap dish on a shelf under the kitchen sink (Beecher 1848, 318). Regular washing and increased attention to personal and household cleanliness came to be included among the fundamentals of gentility, and better cleaning habits were a first step to moving up the social scale (Bushman and Bushman 1988).

In the 1830s, activities of everyday housekeeping began to be codified. Gentility manuals delivered the latest rules of domestic refinement. Guidelines prescribing the exact manners to treat dirt and waste and administer housekeeping were published, at first assuming that the work would be done by servants. With the rise of the middle class, now without paid labor, and the advent of domestic cleaning equipment, rules of domestic cleanliness were increasingly targeted at housewives. This was the beginning of a long-lasting elaboration of cleaning equipment for self and house to facilitate the management of waste, to remove waste in the fastest and most efficient way, to keep clean. Architects began focusing their attention on new spatial properties to enable light and the flow of fresh air into the dwelling—windows in every room, larger spaces inside and between buildings. Starting in the 1840s more householders introduced water into the kitchen. Some installed a hand pump at the kitchen sink, with water drawn from a backyard cistern fed by a conduit leading from the house's roof. Others used a single faucet, usually located in the yard. In his 1842 edition of *Cottage Residences,* Andrew Jackson

Downing, the most notable contemporary authority on landscape and architectural design, still referred to the kitchen as "offensive in the matter of sound, sight, and smells" and recommended its removal from the social and sleeping functions of the dwelling to special house projections and wings or to the basement (1967, 4).

Cottage Residences devoted several pages to the bathroom, water closet, cisterns, water pipes, and proper drains for the kitchen and basement and considered the bathroom "a most desirable feature in all our houses" (5). During the early decades of the century, however, indoor plumbing and water fixtures were still the prerogatives of the rich. The washstand continued its service in middle-class homes and was enlarged to the size of a small table, with a marble top and wooden or marble stand; it became an elaborate cabinet. Washstand basins were not yet sunk into the tabletop. This furniture was called a "lavatory" and survived until running water reached the bedroom.

To address groundwater contamination, alleviate odors, and make better use of dung, the cesspool-free facility was often recommended. The earth closet—a hopper filled with fine dry earth, charcoal, or ashes that drop when a lever is pulled after each load—turned the excrement into fine compost. It was invented by Englishman Reverend Henry Moule in the early 1800s and was promoted a century later as safer, more hygienic, and more productive than the water closet. But the earth closet was rejected in favor of the water closet as soon as municipal water and wastewater systems were commonly available.

The water closet, a fifteenth-century English invention, was introduced and patented in the United States in the 1830s.[7] At first only a few water closets were used, and these were often pan closets served with water through a valve on a supply pipe.[8] In *Cottage Residences,* Downing features a water closet connected to an elevated cistern with a pulling string. Although Downing described such a mechanism as essential, it would not be feasible for some time (1967, 5). Early water closets were built in special projections off the main building and connected to it by a corridor, or they abutted the house's wall and were entered from the outside or through a porch or colonnade. They drained into a stream below, a street gutter, or a hole dug underneath or close to the building foundation. The early-nineteenth-century water closet was greeted with suspicion. Fear of noxious gases in the house was a main deterrent in integrating the closet within the house. Many privies were still located in outbuildings in the rear yard even in large and expensive New York residences. They were often connected to the house by an attractive wooden colonnade or trellis (Stone 1979, 300) (fig. 2.2). The distance of the privy from the house was carefully calculated to avoid the offensive odors yet minimize the walk, a consideration especially on cold winter nights. Some owners challenged the standard, attaching the privy to their dwelling's outside wall or connecting it with a porch to a side or back door.

The yard continued to house a vegetable garden and animals and to serve multiple purposes, though service and cleaning functions were now confined to the backyard, whether in the open or in an enclosed structure. In *A Treatise on Domestic Economy,* Catharine Beecher, a noted authority on domestic science, proposed to connect the kitchen to a multipurpose wooden structure with "back door accommodations" (1843, 276) (fig. 2.3). These consisted of water, waste, and cleaning fix-tures and spaces: an enclosed bathing space with a bathtub; two privy compartments, each with a window; and a designated area for firewood, a specialized bin for coal, and a brick receiver for ashes. Beecher's in-structions and accompanying illustration for a front yard depict an or-namental arrangement with clumps of trees and flowers. The backyard with its utilitarian functions is mostly hidden, while a small structure corresponding with the style of the house, possibly a privy or a garden toolshed, is visible. Her recommendations for the care of yards and gar-dens includes using mixed ashes and sandy soil for optimum texture and moisture and horse dung to fertilize the soil (331).

The period gardens of the well-to-do began displaying new bound-aries that further expressed distinctions between work and leisure. The leisure garden with its irregular lines surrounded the main dwelling, while a geometrically laid-out kitchen garden (about a third to a fifth of the yard area) was usually tucked at the rear or side of the lot. A trellis covered with vines often separated the gardens. A service or work area along with a stable, stable yard, and carriage house was commonly lo-cated near the kitchen garden. Many of Andrew Jackson Downing's gar-den plans included an outbuilding containing two privy compartments, connected to the house by a short path. Referring to a diagram, Down-ing recommended "concealing all objects which would not add to the beauty of the scene by an irregular plantation, as for example, the fence of the kitchen garden at *m,* or the outbuildings at *d,* . . . and in con-trasting these plantations by broad open glades of turf, *n*" (1967, 68) (fig. 2.4). Likewise, the underground system of water and drains—well, cistern, water pipes, cesspool, and drain pipes—is mentioned in the writings but not shown in the ground plans or drawings.

Early in the century the presence of waste and dirt was defined and used to project class status. Toward midcentury, a new element—convenience—became a major factor in how people related to waste and cleanliness. Both Downing and Beecher linked architectural and domestic improvements, particularly air, light, running water, and cleanliness, to ideas of convenience and comfort. These in turn were directly tied to family happiness and, eventually, to national progress. The more practical Beecher emphasized conveniences and making household labor easy and economical, whereas Downing was also concerned with good taste. Especially in the crowded city, where municipal response to worsening health conditions was still inadequate, waste ceased to be a mere nuisance or a matter of class: it became the impetus for a private sanitary reform movement. The principle of convenience came to serve as one of the most important elements of the reform agenda.

Containing waste (1840s–1870s): Waste and comfort

Toward midcentury, people's attention turned to the physical conditions of the city, particularly to miasmic clouds emanating from various filthy sources (such as swamps, cesspools, and slaughterhouses) and the lack of ventilation and pure water.[9] This concern was based on a new popular theory, the filth theory, which refuted claims that epidemics and infectious diseases were caused by a direct contact with persons carrying the illness. But it was the house, the domestic setting of the family, rather than the city where most improvements took place. The desire for convenience—anything or any arrangement that eased and saved domestic labor, safeguarded health, reduced dependence on servants, and improved home life—significantly altered the material condition of middle-class homes. A new interest in personal health gave rise to a privately based sanitary reform movement, later to be replaced by publicly based sanitation reform. Underlying the private movement was the belief that physical space and sanitation, individual character, and national progress were directly connected (Ogle 1993).

In the domestic realm, new codes of sanitation and comfort focusing on proper ventilation and spatial arrangement further increased the separation and segregation of spaces and functions, differentiating work, service, social, and private areas (Canadian Centre for Architecture 1991). Advertisements enticed housewives to take part in the growing consumer culture and buy household labor-saving devices and cleaning materials. As municipalities began building water systems, city dwellers began installing plumbing fixtures in record number. Plumbing was brought into the homes of the middle class.[10] When municipal running water was not available, householders used private sources: water tanks, cisterns and pumps, and, in remote rural locations, nearby springs and rivers. From 1840 on, builders' manuals and architectural texts extensively discussed how to install and use water fixtures.

Architectural plans of the time still featured cheap houses without indoor water closets or bathrooms, but the more expensive homes had one or two of the new features, usually at a corner of the house, at the back of a stairway, or attached to the house in a room entered from the outside. The kitchen appeared at the rear of the house and in some cases in the basement adjoining other storage and service functions (Brown 1851; Vaux 1857).[11] Waste and storage compartments were still attached to the house and accessed from the kitchen or the outside. Domestic conveniences included sink rooms, dumbwaiters, rotary pumps, piped water, hot water, and plumbing fixtures, as well as bathing rooms and water closets (Ogle 1993).

With pipes and running water in the house, the wealthy moved the washstand from the bedroom into a small "washroom," previously a wardrobe or dressing room where clothes had been changed and mended. The washroom represented a new form of privacy. For the first time, cleanliness became associated with a private space (Vigarello 1988, 220). The use of washrooms was extensive among the well-to-do in these years. Women, in particular, enjoyed entering alone and locking the door behind. That locus of privacy was also a source of family pride and a means to impress guests. Americans took pleasure in their elaborate fixtures—plumbing fixtures surrounded by gleaming marble and polished wood—but they also feared that these appliances were a source of the dread diseases still rampant. Faulty plumbing systems in homes were common, giving rise to a new field: sanitary engineering.

The annoyance of privies was yet unresolved, as many still dwelled at the far corner of the yard. Architectural plans from 1869 show in-house privies—supplied with cistern water flowing from an attic tank fed by water from wells, springs, or tanks filled with rainwater—that carried waste into an adjacent cesspool (Ogle 1993, 54). Respected architects and landscape architects attempted to design improvements to this unglamorous facility. Calvert Vaux usually proposed a water closet abutting the house or connected to it by a covered way. He claimed that "a water closet, or its equivalent, is an absolute necessity in any house that is proposed to be a convenient and agreeable residence; but as there is sometimes a difficulty, and always an expense, in arranging a regular water-closet, it is desirable to invent some simple plan, which shall approximate to its advantages at little cost. . . . [W]hen we consider the troublesome, unhealthy, indelicate, ugly effect of these outbuildings, as usually constructed, it is certainly worthwhile to consider this simple plan which is applicable to any house in any situation, and can hardly get out of order practically" (1857, 47–48). Vaux proposed a water-flushed privy arrangement; it included a pipe running from the roof and carrying runoff to an enclosed tank, which drained into a cesspool at a distance. A separate cesspool for sink and laundry wastewater was recommended. His *Villas and Cottages* also included two proposals for outbuildings that resemble ornate garden pavilions; each contained two privy spaces on one side and a toolshed or garden seat on the other.

Vaux's former partner Frederick Law Olmsted also was involved with this mundane matter. He revised the design of earth closets for Riverside, Illinois, and installed four of them in private back lots (Schuyler and Censer 1992, 555). Olmsted believed that earth closets were more sanitary than water closets. The 1868 Riverside plan embraced the concerns of sanitarians and included careful provisions for water supply, underground sewage, and land drainage, as well as larger spaces between houses and ample greenery. Olmsted was appointed in 1861 as the executive secretary of the United States Sanitary Commission, an agency that was a training ground for many prominent sanitarians, and he was extensively involved with hygiene and sanitation matters. In particular, he emphasized the critical connections between waste, sanitation, and landscape design.

Midcentury yards continued the shift from utilitarian to ornamental. In particular, businesspeople settling at the urban fringe distinguished themselves from both the farmer and the city dweller through yards with cared-for, high-maintenance, ornamental plants and curvilinear forms (Stilgoe 1988). Grounds plans of the time rarely showed utilitarian features, such as waste structures and water systems. A fenced-off or enclosed area was designated to contain of waste and messy work. It was labeled "yard" or "kitchen yard," possibly a remnant of the rural "door yard," an area at the side or rear of the farmhouse that functioned as the kitchen and household labor space. Vaux described a "yard" as "a space large enough for a cart to get to the garden with manure, etc." (1857, 202). The architect William Brown featured a "kitchen yard" similar to Beecher's back door accommodation space in his plans. It was an enclosed space, attached to the house's outside wall, and contained an ash pit and a water closet (1851, 60).

During this time, the amount of household solid waste was still relatively small. Housekeeping manuals preached frugality, and most refuse matter had an exchange value—food was sold in bulk, with containers redeemed or reused. Organized garbage collection, typically handled by private companies, began servicing well-to-do neighborhoods. Metal containers, which proved easy to handle and clean, were conspicuously placed along streets and back alleys; they were not yet standardized. Nevertheless, the lack of proper municipal sewer systems and the inability of cesspools to cope with increased water use promoted new epidemics and filth-filled streets.

In the last quarter of the nineteenth century the private sanitary reform movement was undercut by bacteria-based health theories and major technological improvements in sanitation. Many cities began paving streets, installing underground sewers, and establishing garbage collection services. With some paved surfaces and underground systems in place, the landscape of streets and yards changed dramatically, and waste was more easily removed and directed to designated sites. Waste and dirt ceased to be mere elements of class and convenience; they were rationalized as health threats.

Diverting waste (1870s–1890s): Waste and health

Louis Pasteur's discoveries that diseases are transmitted by infectious germs brought material waste, as disease-ridden host, to the center of attention. The germ theory ultimately provided a rationale for the connection between cleanliness and disease prevention. Citizen reform movements spearheaded the attack, mobilizing an unprecedented crusade on rotten garbage, decaying food, and human and animal waste. Cities appointed public health boards, established public works departments, and enacted major laws to control pollution.[12] Sanitarians began to formulate plans to remove waste rapidly, to render it invisible and odorless, and to preserve the population from any contact with it. Sewage systems were built and buried underground, and dumps were pushed to the margins of the city. The bathroom and water closet were widely adopted, and streets and yards were cleared of human waste. At the same time that waste was domesticated, the responsibility for waste management was transferred to the public. Interest in domestic sanitation reached its apex at the International Health Exhibition in London in 1884, named "The Healthieries," which was visited by American sanitarians and public officials (Canadian Centre for Architecture 1991) (fig. 2.5). During this time, doctors formulated architectural space on medical principles. Houses were likened to bodies.

From 1880 on, the little washroom attached to the bedroom, or in some cases an entire small bedroom, was transformed into the mechanized bathroom. New houses began designating specialized rooms as bathrooms. As the bedroom extended itself and its furnishings into the washroom, the washstand was transformed from furniture into fixture. The sink was sunk into the marble top, and water pipes were encased in and fixed to the wall. Until the end of the nineteenth century, lavatories came in expensive materials and highly decorative designs. The availability of cast-iron lavatories lowered the price to within reach of the masses.

Bathtubs appeared in growing numbers and in a great range of varieties and styles. There were pricey porcelain and more affordable enamel-covered cast-iron or fireclay tubs. There were even lead traveling baths with straps and collapsible legs and waterproof papier-mâché baths of small and lightweight design. New York's J. L. Mott Iron Works catalog of 1888 featured more than 275 lavishly illustrated pages replete with everything from dolphin-shaped commodes to claw-foot bathtubs and imported French bidets. Most city dwellers, however, did not have a bathtub until after the 1900s (Wright 1960, 165), and only about one-quarter of urban householders had water closets, while the remainder still depended on privies (Tarr 1984, 231).

In the standard New York house of the 1880s, sanitary fixtures were supplied by gravity flow from an attic storage tank, to which water was usually raised by means of a hand pump. Textbooks with illustrations showing cross sections through houses alerted people to the faults of plumbing and installation designs. The first issue of *Plumber and Sanitary Engineer* magazine (1879), later called *Sanitary Engineer*, published diagrams of "correct" drainage and articles by experts.[13] Women and plumbers were designated the guardians of domestic sanitation. However, it was the water-carriage technology—the sewerage—that had a major impact on people's lives as it relieved them of a sensory nuisance and labor-intensive work; more important, it transformed a private matter into a public concern. Private waste became a public issue and a collective problem. With sewers in place, the kitchen and the bathroom in the modern house became the domestic places for the instant elimination and transfer of waste, releasing the yard and the individual from major waste-related functions. Journals featured modern bathroom designs with porcelain tiles and freestanding fixtures with exposed plumbing pipes, emphasizing the ease of cleaning and fixing them (fig. 2.6).[14] Soap entered the national consciousness and became a symbol of status and bodily and domestic hygiene. Kitchen design in the late nineteenth century stressed convenience, economy, and hygiene. Catharine Beecher's housekeeping manual recommended whitewashed kitchen walls "to promote a neat look and fresh air" (1873, 93). Linoleum for kitchen flooring became available and popular in the 1870s.

During that period, garbage, rubbish, and ash piles still dotted the sides of streets. Annoyed yet determined pedestrians began to organize civil protests. The City Beautiful movement, which swept the nation in the 1890s, provided the rhetoric equating the elimination of pollution with an idealized city aesthetic (Melosi 1982). Women's involvement in refuse reform and city housekeeping was seen as an extension of their domestic role. As a statement issued by the 1884 New York Conference of the Ladies' Health Protective Association (LHPA) proclaimed, "even if dirt were not the unsanitary and dangerous thing we know that it is, its unsightliness and repulsiveness are so great, that no other reason than the superior beauty of cleanliness should be required to make the citizens of New York, through their vested authorities, quite willing to appropriate whatever sum may be necessary in order to give to themselves and to their wives and daughters that outside neatness, cleanliness and freshness, which are the natural compliment and completion of inside order and daintiness, and which are to the feminine taste and perception simply indispensable, not only to comfort but to self-respect" (Melosi 1980, 113).[15] Consequently, public works departments busied themselves with street paving and cleaning in a major project to sanitize public places.[16]

The back alley continued to maintain its service and storage functions, including refuse collection, the passage of utility wires, and access to the backyard and the newly added freestanding garage. The exodus of the middle class from city centers to the emerging class-segregated suburbs accelerated. The new suburbs imitated the Riverside model of ample space and greenery, wide curving roads for leisure drives, and disguised utility and waste systems.

Toward the end of the century, science, technological advances, and higher standards of living cleared the way for higher standards of cleanliness. Public demand for a healthier, cleaner environment in the city was coupled with a demand for beauty. Cleanliness became clearly and intimately connected with aesthetics.

Clean is beautiful (1890s–1910s): Waste and aesthetics

From the end of the nineteenth century onward, gentility, convenience, health, and a new element, aesthetics, would converge to form a nexus of values at the heart of Americans' attitudes toward waste. While indoor spaces—kitchen and bathroom—gained firmer possessions of their domestic sanitary functions, the yard was transformed into an aesthetic leisure ground.

Kitchen design of the 1890s continued to emphasize convenience, economy, and hygiene. The better-equipped and easy-to-maintain kitchen reduced dependence on servant work, and the space was elevated from the basement. The majority of kitchens were still at the rear of houses, but they were now built on the main floor. The kitchen's character shifted from a large, lighted room to a smaller, compact, sanitary, laboratory-like workplace. The laboratory image of the turn-of-the-century kitchen epitomized Frederick Winslow Taylor's desire to turn the kitchen into an organized work space and manage all housework factory-style, in order to impose more efficient work patterns. Taylor's scientific management sought to eliminate waste of time, effort, space, and money. As cultural scholar Helen Molesworth points out: "In this context, the double valence of the word 'waste' cannot be overlooked. The desire to eliminate waste, both literally and rhetorically, is in part a desire to consolidate a culture's understanding of cleanliness. That the kitchen should become a site for Taylorization suggests the doubleness of the kitchen's simultaneous roles as both waste producer (garbage, dirty dishes, etc.) and waste manager (work process, trash removal, cleaning)" (1997, 84).

Likewise, in their analysis of the evolution of domestic waste treatment spaces, interior designers Ellen Lupton and Abbott Miller observe that "The kitchen became a site not only for preparing food but for directing household consumption at large; the kitchen door is a chief entryway for purchased goods, and the main exit point for vegetable parings, empty packages, leftover meals, outmoded appliances, and other discarded products" (1992, 1). The kitchen's floor and walls were made washable, easy to clean; tile or linoleum was used for floors and white tile, vitrified brick, or specially treated plaster for walls. Water and drainage pipes were left exposed "so germs and dirt cannot collect in hidden areas" (Plante 1995, 163).

The new expectations of the mechanized and hyperclean kitchen and home led to a burgeoning market for household products. Increased consumption and the growing use of newspaper advertisements at the turn of the century created increasing amounts and new types of waste. People started burning their trash and simply threw away more. As for the garbage can itself, people used whatever barrels and boxes were in hand. The social historian Susan Strasser tells us that "Neither Sears nor Montgomery Ward carried any household products designed and marketed specifically for holding or disposing of garbage; no trash barrels, garbage cans, or waste paper baskets" (1992, 10).

In the following decades, books and magazines began publishing specific, practical instructions for dealing with domestic waste. In *The Efficient Kitchen* (first published in 1914), Georgie Child provided comprehensive suggestions. Householders were encouraged to leave the kitchen free of any garbage pail or basket. The garbage container was to be placed outside the back door, buried underground or built into an outside wall, and the drained garbage should be wrapped in a newspaper or a grocery bag. She advised, "For those who have coal stoves or furnaces the simplest plan of garbage disposal is to burn it up" (Child 1925, 171). Wealthier householders could use the costly gas-fired "destroyer," placed beside the kitchen range, to burn their garbage. A special outdoor refuse burner could be used in the summer. Ashes were thrown into the usual garbage can or buried in the garden. Child recommended that "Every householder ought to have a strong galvanized garbage can" (Child 1925, 171). Another suggestion was to use an agateware pail, which was kept in a box with a hinged cover and sunk underground. For those who could afford it, she suggested a special cast-iron container with a hinged top designed to be installed in the ground outside. The garbage receptacle was placed inside, and its cover could be opened with a foot-operated spring. According to Child, "the chief cause of unsightly back yards is failure to plan intelligently for the care and disposal of these various kinds of waste" (174).

At the turn of the century, bathrooms were changed from furnished and elegant rooms to simple, practical, and sanitary lab-like spaces. Woodwork, curtains, and carpets disappeared in favor of white enamel, tiles, and marble. The bathroom shrank into a small room with all plumbing fixtures on one wall. Sinks came in one-piece whiteware or glazed fireclay, complete with flat tops and recesses for soap. Sinks, with or without a pedestal, were stripped of decorations; joints were left exposed for easier cleaning.

The receptacles of dirt—sinks, tubs, and toilets—embodied the modern idea of private hygiene. These loaded dialectical sites were accepted in the private domain only in their immaculate cleanliness. They collected and channeled dirt down the dark hole of the drain, yet remained pristine and clean. Shining metal, enameled cast-iron, or porcelain, their smooth and bright appearance contrasted with the sullied content they processed. These receptacles, each in its own way, were also loaded signifiers of the body. The bathroom underwent a process of rigid standardization and efficient spatial configuration, gathering all three fixtures into a unified private zone. Sanitary fixtures at reduced price became affordable to the masses. As with the kitchen, claim Lupton and Miller, "The bathroom became a laboratory for the management of biological waste, from urine and feces to hair, perspiration, dead skin, bad breath, finger nails, and other bodily excretions" (1992, 1).

The yard, like the house, played an important role in the newly formulated aesthetics of cleanliness. Major shifts were visible in both front and back yards: larger and more open space, increased tidiness, and fewer boundaries and margins. The 1890s witnessed the eradication of the front yard fence and the area's spatial transformation into a continuous, manicured front lawn in middle-class neighborhoods. The common American backyard gradually joined and underwent similar changes, but not before it rid itself of waste functions. The removal of cesspools and outdoor privies was the most significant element in its transformation.[17] What remained—trashcans, compost piles, and furnace ashes—slowly became a source of shame. Subsequently, these waste materials and objects were relegated to a specialized service area or to marginal side strips, often immediately next to the garage. Much of the residential yard space was freed up for new social functions and family recreation; the front lawn extended to the back creating a unified green carpet around the house.

The desire to disguise utilitarian functions and to obscure weeds and evidence of decay to achieve neatness and a unified composition in the garden began to dominate public taste, as can be seen in the writings of popular house and garden magazines and books of the turn of the twentieth century. Civic improvement organizations urged small-lot householders to contribute to civic beauty by planting flowers, vines, and grass in their backyards in place of the waste and rubbish then customarily dumped there (Petersen 1979, 95) (fig. 2.7). Simultaneously, attitudes toward back alleys changed. As the neighborhood spirit awakened to a new sense of beauty and orderliness, alleys appeared to be a discordant element.

Questions about the alley began to be raised in the early twentieth century, as its utilitarian function diminished and utilities were buried under the street pavement. Worries about aesthetics and maintenance contributed significantly to its shrinking popularity and eventual elimination. As streets and yards were cleaned up, alleys became intolerable and took on a fearsome image—dirty, run-down, rat-infested, crime-ridden (fig. 2.8). Some cities began cleaning and paving back alleys, but these improvements did little to change the attitude in residential neighborhoods (Martin 1995, 65). Their hostility was made clear in the 1912 City Club of Chicago design competition for a residential quarter. The majority of entrants proposed neighborhoods with no back alleys, and the winning designs and jury comments argued boldly against the service alley on economic, practical, and aesthetic grounds.[18] In addition, a heightened sensibility regarding the domain of the individual was expressed in a desire for a truly private rear yard. Consequently, during the first three decades of the century, alleys appeared less and less often in the platting of new neighborhoods.

Organized municipal or privately contracted garbage collection and street-cleaning services in large cities, generally inefficient prior to 1900, were gradually perfected and systematized during the first two decades of the twentieth century. After study, authorities announced regulations on the placement of garbage cans. Professional journals were filled with discussions of can size and construction, dump wagons and collection trucks, and effective management. Municipalities passed ordinances requiring that garbage be drained and wrapped before being placed in cans (Ogle 1988, 26). Child mentions special garbage bags that fit into a standardized can; these were used in the Boston vicinity to ensure its easy handling and cleaning (Child 1925, 171). City officials and health experts stressed the importance of public cooperation by distributing printed copies of collection rules and by fining reluctant, uncooperative households or refusing to collect their garbage. Because garbage was often sold as animal feed, some cities had special requirements forbidding the disposal in the cans of anything that could not be fed to animals. Many cities mandated separate cans for ashes, trash, and garbage and sold the materials to private recycling companies.

During World War I (and later during World War II as well) an attitude of conservation and thrift temporarily took hold. The federal government launched recycling programs with the motto "Don't waste waste—save it."[19] After the war Americans quickly reverted to wastefulness made possible by unprecedented affluence and encouraged by a booming consumer economy. For manufacturers, waste became profit; for middle-class America, it meant efficiency, expediency, and leisure.

Sanitizing spaces (1910s–1940s): Waste and efficiency / leisure

Population growth, mass production and consumption, affluence, and suburban expansion were the major causes of increasing volumes of household solid waste. New kinds of waste were created in support of market growth—excessive packaging, printed advertisements, sales catalogs, built-in obsolescence, and disposable products. Dirt was seen as an obstacle to progress, while waste, now comprising the remainders of all used commodities, was seen as the inherent ingredient of progress. The 1920s elevated the garbage of consumer culture into a form of positive production, valuing the destruction and replacement of objects as a pleasurable and socially useful act. Products were designed for a clean, throwaway culture (Lupton and Miller 1992). The new standard of living drew a blow to the discipline of individual frugality and reuse practice (with brief exceptions during the two world war periods; see chapter 4).

In *Waste and Want,* Susan Strasser incisively describes the social processes then at work. Wastefulness, Strasser explains, was endowed with practical and symbolic powers. Disposable products made cleanliness easy, required no labor, and released people from the need to care for and maintain things. Wasting was considered a good thing, a key to convenience, freedom, and a prosperous economy, while recycling and reuse came to be associated with poverty and backwardness. Strasser also points out that disposability was politicized, signifying the freedom of capitalism as opposed to the bondage of communism. The ideas of waste, abundance, and American democracy were seen as interrelated (Strasser

1999, 226). Disposability became the new resource ethic; waste, an essential ingredient for efficiency and expediency. The new standards for personal and domestic hygiene exceeded the demands of comfort, gentility, health, and aesthetics; they became means to save labor and increase leisure time.

The house plan exhibited consequences of the new ethic. A trend toward multiple bathrooms began around the 1920s, as the bathroom reached the height of privacy, establishing itself as an appendage to the bedroom. The three-fixture bathroom unit now reached its ultimate standardization.[20] The cell-like shower cabin, which was invented for public showers and was hygienic and cheap, also found its way into middle-class bathrooms. But clearly, the tub was perceived as "more relaxing, luxurious and feminine in usage than the shower" (Kira 1976, 20). In the 1940s companies used mail-order catalogs to sell the full set of fixtures: tub, basin, and toilet. Conversely, the kitchen was rescued from its isolation and opened up as shared domestic space, no longer a nuisance producing bad smells and messiness. In the 1930s, the "living kitchen" concept emerged, combining the kitchen and dining room.

Domestic furniture and fixture designs took on a new form, expressive of the new ideals of efficiency and expediency. According to Lupton and Miller, the first response of modern design's imperative for cleanliness during the 1920s is expressed in enclosure and streamlined design, as evidenced by the seamless, aerodynamic form of many items and in the overall transformation of the basin, stove, and bathtub from bulky, freestanding items to continuous kitchen or bathroom counters and built-in cabinets. The built-in design united all fixtures and furniture with a smooth, enveloping surface without decorations; mechanical joints were minimized for efficient, expedient, and easy cleaning. According to Lupton and Miller, "Streamlining performed a surreal conflation of the organic and the mechanical: its seamless skins are fluidly curved yet rigidly impervious to dirt and moisture" (1992, 2). Everything from floors to walls to objects was designed to be easily washable and quickly shed dirt (fig. 2.9). The answer for growing quantities of household solid waste was a larger outdoor garbage container. Little space or attention was given to the storage and processing of waste within the house. Lupton and Miller explain that "The placement of garbage cans was not seriously considered during the evolution of the continuous kitchen. It was repulsive to actually picture waste; period views rarely show the can at all, which is assumed to be snugly stashed beneath the kitchen sink" (1992, 72). The continuous kitchen concept, therefore, created the illusion that waste simply disappeared or never existed. Kitchen idea books, too, did not touch on the subject of waste.

But new prominence was given to the garage, which gradually joined the kitchen and bathroom as a domestic service space, storage for the transaction of waste. Originally a converted horse barn or outbuilding abutting the back alley, the garage became in the 1920s an essential adjunct to the dwelling, though still without direct access to it.[21] A decade later the garage was attached to the house, directly connected to the kitchen and oriented to and accessed from the front street (Jackson 1980). In 1927 some cities started limiting waste collection services to front streets alone.

The yard continued to establish its leisure and recreational roles and to lose its connection to production and work. "Outdoor living rooms"—terraces, grills, patios, and sun porches—were increasingly featured in home and garden magazines. The family moved outside, as waste and storage space moved in. Growing suburbanization and larger grassy yards created a new type of waste, yard waste, which was efficiently displaced to public landfills. And as the need for a productive yard disappeared—no domestic animals or vegetable garden—compost piles were scarcely needed.

By the 1940s, most urban streets were paved and municipalities were enjoying technological advances in equipment. All garbage collection vehicles were enclosed and motorized.[22] Many cities replaced open dumps with sanitary landfills and experimented with various disposal technologies. Waste was now perceived primarily as a technical problem that should be left to sanitary engineers with specialized training. Consumers (or waste producers) became increasingly distanced from waste, unaware of and oblivious to its scope and consequences.

Out of sight, out of mind (1940s–1970s): Waste and obliviousness

The prosperity and dramatic metropolitan growth that followed World War II were the main factors driving a huge increase in the volume and complexity of consumer waste. The size of garbage containers increased, as did the number of collection days. Household waste contained disposable and excess packaging made of synthetics, paper, and plastics, as well as environmentally harmful materials. Wasteful habits permeated American society, but waste itself passed under people's eyes virtually unnoticed; placed in its specified container, it disappeared the next day.

Very little changed in the kitchen and bathroom in response to the growing amount of waste. New kitchens introduced more storage space for consumer products and less for consumer waste. The kitchen incorporated two new appliances for managing household waste: the trash compactor and the in-sink garbage disposal. While the trash compactor, which compresses household waste to 10 to 20 percent of its original volume, has not attained widespread acceptance, the in-sink disposal became quite popular starting in the 1950s. It diverts some of the organic solid waste into the sewage system, removing it in one step from the kitchen where it originates.[23] Now directly connected to the garage, the kitchen became the primary entry and exit for the house's inhabitants, and the garage an intermediate space for transactions of consumer goods and waste. The garage grew spacious and took on the household work, service, and storage functions previously associated with the door yard and backyard. Utility storage, including garden tools and bicycles, moved into the garage and the remaining marginal spaces of the yard disappeared.

The last leftover source of shame, the garbage can, was still commonly placed outside, next to the one-door garage. The twenty-gallon metal container of the early twentieth century—easily bent, liable to rust, and noisy—was slowly replaced by thirty- to thirty-five-gallon plastic containers. Disposable paper and plastic bags joined the scene in the 1960s, and some local governments mandated their use. The major change, however, was the enlarged, two-door garage, which became common in affluent suburbs of the 1970s and enabled garbage containers to be placed inside the garage. With the residential alley officially banished from neighborhoods built after World War II, refuse collection was now relegated solely to the front of the house. Municipalities enforced new regulations pertaining to the location and times of putting out garbage cans and bags in the streets, limiting collection to early morning hours.

Once the garbage can was removed, the backyard lost the final obstacle to its total transformation into a recreation area for the family—a place purely of play, not work. Garden magazines urged readers to "get more living" out of the yard. More lawn area and large outdoor patios and terraces, as well as open structures for eating, barbecuing, and resting, dominated the yard. A service area, if introduced at all, was very small and attached to the garage or tucked at the rear of the yard, usually enclosed by a fence "to hide the garden work" (Sunset 1952, 30). Commercial fertilizers rendered composting impractical and old-fashioned. By 1970, yard waste, especially grass clippings, grew to 10 percent of total municipal solid waste.

The municipal response to increasing volumes of garbage included upgrading and adopting of large-capacity trucks. The "Godzilla" garbage truck, with a loading arm and mechanical compactor for ninety- to three-hundred-gallon containers, paved the way for a variety of mechanical collection trucks that further encouraged larger waste containers and more wasteful habits (Melosi 1981, 236). Disposal practices moved waste away from residential spaces and into remote and disguised public facilities, but it struck back in the form of air, water, and land pollution. In the 1960s, a sense of environmental crisis was the impetus behind a major conservation movement and new laws. The 1965 Solid Waste Disposal Act was the first recognition of refuse as a national issue in the United States. A national crusade to control and eliminate waste was declared, marked in 1970 by the first Earth Day and the formation of the Environmental Protection Agency (EPA). Soon after, waste came to be viewed as a resource and recycling a necessary and politically unassailable environmental ethic.

Waste recycled (1970s–1990s): Waste and resource

More than any other issue, solid waste became a symbol of the environmental crisis, and in response, a variety of public and private recycling programs began. They were organized by environmental enthusiasts and boosted by monetary incentives, such as laws in many states that required a deposit on glass bottles and metal cans. By 1979 thousands of nonprofit and small commercial recyclable collection centers existed (Hoy and Robinson 1979, 6). In the early 1990s, more than 140 laws that bear on recycling were passed in thirty-eight states, and a national lobbying group set a national goal of achieving a 75 percent recycling rate for municipal solid waste by the year 2000. In 1994, more than 101 million Americans were served by curbside recycling programs (Steuteville 1994, 46). Mandatory recycling came into effect in many large and small municipalities. Recycling became a symbolic act of caring, of saving raw material and reducing pollution, of participating in something important. Some people even considered recycling an effective act to counter capitalistic consumerism.

The new reality was slowly integrated into the house, yard, and street. As early as 1981, *Sunset* magazine displayed ideas for kitchens with special custom-made fixtures for waste management and recycling, including a plywood portable box for recyclables near the kitchen door and a tip-out waste bin or dual-duty drawer below a cutting board for food scraps and recyclables. Evidence of a broad-based awareness of this shift became apparent only later in the decade. Patricia Poore, editor of *Garbage* magazine, whose first issue in 1989 was published to great acclaim, claimed that "we are on the brink of the next design revolution" (1990, 35).

Several design and waste management books and journals began exhibiting new ideas for kitchens and basements (J. Poore 1989; P. Poore 1990; Brown 1989). The new kitchen was designed to accommodate bulk packaging, food waste composting, recyclables separation, water conservation, and energy efficiency. Specific changes included built-in composters, vented cabinets for recyclables, chutes that carry used materials to storage into separate bins in the cellar or yard, recycling bins tucked under the countertop, wastewater processing equipment in the basement, and recycled graywater storage in the yard. The Smart Kitchen, a design by David Goldbeck, rejected the idea of continuous cabinetry in favor of freestanding countertops of adjustable height and a mix of open and closed storage. Vegetable scraps are dropped into a chute connecting the countertop with a can underneath. The can, accessed from the outside, empties into a compost pile. Recycling bins are incorporated into the cabinet space on a pullout platform or wheeled cart that can be rolled directly to the curbside (Goldbeck 1989). In high-rise buildings, residents could throw sorted recyclables in separate chutes that directed the material to large bins in the basement. Automated recycling equipment was tested. Residents placed recyclable trash in chutes and pushed a button to activate separate receptacles revolving at the bottom (*Biocycle* 1994). Such installations advanced the transformation of the house from a unit that only consumes products to a unit that processes consumer waste as well. One design concept, William Stumpf's *Metabolic House*, shows the house as a highly articulated garbage center (Brown 1989) (fig. 2.10).

New designs for bathrooms showed a trend toward larger leisure or recreation rooms. Bathrooms were featured with skylights and plants, hot tubs and whirlpools (*Interior Design* 1994). A modified toilet model that works with significantly less water was introduced.[24] Massachusetts has required the "ultra-low-flush" toilets (using 1.6 gallons per flush) for all new construction since 1989 (Kourik 1990, 22). In an effort to transform the house and yard into a self-sufficient unit, various technologies and systems that handle household effluent were tested. Researchers developed several odorless and feasible options to avoid the sewer system—composting toilets, chemical toilets, and on-site graywater purification systems, such as reed beds and microbial filter beds (Del Porto and Steinfeld 1999). A modern composting toilet arrived in the United States in the 1970s.[25] The early installations in the 1970s and 1980s suffered from maintenance and structural problems. With better marketing, more accommodating regulations, improvements in the composting mechanism, and progress at overcoming the negative perception, the popularity of composting toilets has increased. Generally, however, most people, especially city dwellers, were content with their water-consumptive model—the only option they have (Tarr 1999, 5).

The yard began assuming a key role in the waste-processing effort, and several old-time fixtures made a modest comeback: compost piles, cisterns, and recycled water irrigation systems. The growing sentiment to reuse and recycle domestic wastewater in the 1990s produced two related terms, "graywater" and "blackwater." The first is the flow from laundry, sink, and bath, the second from the toilet. Each can be treated and reused locally for irrigation or household cleaning. In the 1990s, as disposal prices rose, many states banned yard waste from municipal sanitary landfills, thus encouraging residents to change their behavior. Municipalities followed with ordinances allowing yard waste processing, thus paving the way for the reintroduction of the compost pile and graywater irrigation. Indeed, half of the single-house American households that maintained gardens in the 1990s had a vegetable garden and composted organic and yard wastes, returning their nutrients to the soil in the form of mulches or chipped wood (Girling and Helphand 1994). But many people still preferred that the city or the many private composting sites that then mushroomed handle their yard waste.

Permaculture practice began manifesting radical changes in green enthusiasts' yards across the country. According to Bill Mollison's *Permaculture,* the world is inherently untidy. Neatness, tidiness, uniformity, and straightness, all of which signify social order, are in design or energy terms disordered. "True order," he suggests, "may lie in apparent confusion, . . . Thus the seemingly-wild and naturally-functioning garden of a New Guinea villager is beautifully ordered and in harmony, while the clipped lawns and pruned roses of the pseudo-aristocrat are nature in wild disarray" (1990, 31). Ecologists advocated a yard with complex boundary condition—a mix of grasses, plants, and ground cover—an edge mosaic to increase diversity and productivity. Dead leaves, branches, and even trees commonly removed as nuisances or eyesores constitute ecological havens, hosts for birds and insects as they recharge the soil with nutrients. This ecological argument offers but one way to understand and evaluate how social conventions of waste and cleanliness affect landscapes.

Simultaneously, changes in neighborhood planning were apparent, particularly in the so-called neotraditional communities. Alleys were reestablished, the front yard picket fence reappeared, and the garage was pushed to the back. The designer and planner Peter Calthorpe believed that among other benefits, secondary access (through a back alley) allowed people to undertake messy tasks without offending their neighbors. Philip Langdon also recognized the need for outdoor work space in the yard: "People need places where they can conduct unpolished activities" (Langdon 1994, 157). Architects and planners recognized edges for their capacity to create not only rich niches for organisms but also diverse forms, stimulating landscapes, and enriched human experiences. The "edge effect," a concept known to both ecologists and environmental psychologists, which makes attractive places for organisms and people to locate, was acknowledged. Mollison argues that "more events occur at boundaries than occur elsewhere" (1990, 77).

Recycling bins, stationary or rolling, of all sizes, textures, and colors became commonplace on the porch, in the yard, at the curbside, or by the streetside parking strip. Manufacturers produced dozens of different compost and recycling bins, and some cities even distributed them free or at low cost and established composter training programs for both suburbanites and urbanites. Social approval for the new waste recycling norms clearly was high. The new landscape testified visibly to the changed status of waste. Specialized recycling bins added a colorful, ephemeral quality to the street as well as contributed to a sense of environmental activism, communal responsibility, and good citizenship.

Perceiving waste as a resource was a radical departure from the dominant preceding views. It freed waste from its associations with impurity, disorder, formlessness, and death. It purified it, put it in order, and gave it form and the power of rejuvenation. Seeing waste as a resource began to loosen its connection to class, health, efficiency, leisure, and aesthetics. (In fact, it has created new aesthetics and new leisure patterns.) Indeed, the transformation of waste into what we consider recyclable objects—the separated, orderly, and clean arrangement of recovered material—has stripped some of the negative associations from waste. Growing consumerism and sophisticated electronic technologies have both supported and contradicted this trend.

The smart house and yard (1990s–present)

At the turn of the new millennium, new realizations have surfaced: the recycling euphoria needs more grounding in economic sustainability, and the "green ethic" might be loaded with excessive morality that lacks broad appeal and effectiveness. Waste management has come under the control of a few giant corporations, which also controlled recycling habits and public sentiment. As they claim the label "green corporations," they use (or abuse) the new image for their own profit. At the same time, small, innovative enterprises have begun to offer ideas to deal with excess and waste that make good economic sense and can compete in the market. Research and educational organizations direct their efforts toward innovation in the smart utilization of energy and material. New technologies and high-style sanitary fixtures have been developed with conservation and efficacy in mind. Consequently, decentralized systems and the return of private responsibility for waste management are on the rise. Input and output are maximized. The house and the yard are becoming one integrated unit, working in concert to process waste and produce food and energy. Wastewater and human waste can be converted to high-protein nutrients for urban agriculture and aquaculture. Consequently, the supermodern domestic scene begins to look as follows.

The floor plan of today's model house reflects and facilitates the individual self-fulfillment of all family members. The trend of more and smaller efficient rooms "to imaginary people and activities" is evident (Chapman 1999, 1). Hidden wires network all the smart electronic technologies in the house, connecting them to sources of services, supplies, and information through broadband Internet access. Voice-activated machines and microprocessors and sensors that react to people entering the house and respond to their needs—cleaning, doing laundry, preparing dinner—promise to become the way of the future.

At the turn of the twenty-first century, the kitchen reinforces its status as the house's most important family space. It is turning into a computerized control room for managing consumption. The kitchen-dining space has already grown larger and separate from the living room. Compared to the kitchen of the 1980s, which used only a small part of its input—water, food, energy, and packaging—the new kitchen is growing more efficient. For example, the "smart kitchen" of England's Cranfield University and Electrolux Industrial Design Centre is based on the food preferences and dietary needs of the inhabitants. A "datawall" acts as a kitchen brain that keeps track of food stock and freshness. It is also linked directly to the supermarket to facilitate accurate orders of supplies. The sink is made of a membrane that expands when filled with water, connected to a graywater filter and to a graywater storage tank (Hinte and Bakker 1999, 67). Manufacturers of smart appliances are working on a line of refrigerators that will send e-mail messages to owners regarding needed items.

This smarter kitchen continues its trend toward maximum flexibility with wheeled, independent units. A mobile, stainless steel set of stove, cabinet, and counters in the demonstration kitchen at the Center for Maximum Potential Building Systems (CMPBS) in Austin, Texas, can be rolled to any point in the house and to the outdoor patio for cooking. Its parts can be stored in a closet when the space is needed for a larger crowd.[26] The kitchen becomes a resource for undertaking vermiculture and aquaculture in a greenhouse or sunroom. These home-scale operations process food scraps and other organic waste by using earthworms and fish ponds. Neighborhood ponds can become a small business, and the worm castings are an excellent soil additive. A large utility room attached to the kitchen mediates between it and the garage. Garages, too, are becoming cleaner and more organized, more spatially distinct from the main house and in some cases detached.

Many new houses have a small bathroom attached to each bedroom. The contemporary bathroom is a highly individualized space, reflecting an increased bodily taboo within the family. On the one hand, the bathroom is a place to indulge and celebrate the body; on the other hand, it is a space to monitor the body's health. New accessories, such as massaging shower heads, stylish organizers, and heated toilet seats, find their way into the standard bathroom. The health functions include appliances to monitor weight and blood pressure. Bathrooms of the affluent are large, well-lighted, and complete with whirlpool bath, bar, and digital entertainment center. In the new bathroom, style matches function, and owners show off their bathrooms as a reflection of their personalities. As bathrooms become again places of renewal and regeneration, people spend more time in them. Conversely, at the CMPBS demonstration center, Pliny Fisk has designed a small, flexible bathroom; it contains fixtures that rotate around a central column and can accommodate up to two people privately, using a heavy sliding curtain to separate them.

The privacy of the toilet and the secrecy of its functions have been challenged recently by some artists, who offer different models for our hygiene spaces. Austrian artist Cornelius Kolig, for example, has built a unique residential complex called *Das Paradies* that serves as an instrument for exploring the relationships between home, nature, and the human body—sexuality, excretion, and death. An isolated area at the complex that is shaped as a contemporary basilica features a series of spaces or small temples dedicated to bodily functions—defecation, urination, and ejaculation. Urinals and containers are beautifully crafted and gracefully reveal the products of these functions. In the conservatory of the sun male's urine is led down a sloping cylinder into a reflecting pool, activating concentric ripples in the water and distorting the sun's rays. In another space, the women's urinal covered with yellow canvas directs the liquids to a nearby meadow. Countering centuries of sublimation and elevating the profane to the sacred, Kolig projects his eccentric, counterculture manifesto, returning us consciously to paradise's naïveté (Galfetti 1999, 182; Ritter 1997, 50).

Scientists, concerned with the efficacy and sustainability of the toilet, have developed on-site wastewater treatment systems that separate graywater from blackwater and convert them to useful resources. Graywater fed through a subsurface pipe system and microbial rock plant filter can be used to irrigate planting beds, yards, and greenhouses, especially in southern states. The garden around the house becomes integral to a closed system of cleansing and producing.[27] At the CMPBS research center the toilet is equipped with a composting vessel where solids and liquids are separated; the solids are processed as composting material and the liquids are purified via reed beds and a leachate field. The two reed beds, which also make space for six sealed tanks that store water from the roof, adorn and frame the building entrance. Such decentralized and small-scale facilities are easy to control, cost relatively little, offer water and energy efficiency, and promote local economic development.

The reuse of wastewater has therefore prompted new types of gardens, engineered gardens—"washwater Gardens" for graywater and "wastewater Gardens" for combined effluent (graywater and blackwater). The washwater garden, which cleans and uses graywater, is based on the work of Dr. Alfred Bernhart from the University of Toronto. It relies on aerobic microbial respiration and evapotranspiration. It could be built in or near the house in a variety of forms—a special planted bed, an entire landscape surrounding the house, or indoor planters in an enclosed greenhouse or off a sunny living room (Del Porto and Steinfeld 1999, 181). The roots of flowering plants, such as calla and canna lilies, bulrush, cattail, and elephant ear, break and absorb harmful substances as water flows through them.[28] A sizable operation can even become a profitable flower business. The wastewater garden receives prefiltered and pretreated water for further polishing. Specialized wastewater treatment systems using aerobic and anaerobic digestors that produce methane gas and fertilizer treat the blackwater first and ready it for the garden. The garden is planted with phreatophytes—broadleaf and rainforest plants such as bamboo and reeds. Aerobic microbes digest the remaining waste and transform it into heat, water vapors, and gases. The plant biomass of either yard is harvested and can be fed to fish and composted. Neighborhood aquaculture ponds can serve as an alternative sewage cleansing system (Del Porto and Steinfeld 1999, 183).

Improved composting toilets are making an appearance in the market as they become more socially acceptable. The new models are easier to maintain, more efficient, cheaper, and, most critically, lack smell; they also are stylistically more refined. Service companies offer maintenance, more states are permitting the toilet under new laws, and the EPA is encouraging the formation of district organizations to manage on-site systems. Sludge collected from tanks (sized appropriately for the building) is delivered to and treated in a central community facility, where it is mixed with sawdust or chopped wood and composted.[29] The "smart" composting toilet now coming onto the market uses solid-state sensors and microchips to control the drying and composting process. Another system automatically deposits high-powered composting microbes into the collected waste. Both require the owners to do little. Researchers propose to build composters directly into a building's foundation. Composted products would be removed in roll-away composters or fully packaged in bags to a composting facility or a garden. Management districts would form to oversee the effectiveness of the new system and enjoy federal money that now goes to sewage systems. The use of the composting toilet (both dry and ultra-low-flush) is on the rise, not only because septic tanks are now outlawed in some remote and high groundwater areas where sewers are not feasible or economical, but also because municipalities are weighing the economics of their options. Decentralized on-site systems make highly expensive sewage treatment facilities unnecessary and save large amounts of water that would otherwise be wasted.[30] If the decentralized sewage system trend continues, the private residential landscape of the future would likely be dotted with gardens and greenhouses. These engineered landscapes would be working facilities as well as beautiful places of regeneration, producing fuel, food, fish, and a healthy habitat at the neighborhood scale.

No "primary" element can be singled out from the complex forces that shape our environment; however, the crucial link between waste and private landscapes is obvious. In each period discussed above, people used the idea and material of waste to give order to, make sense of, and better their world.

Contemporary residential landscapes are taking a significant turn. Environments (and products) are designed not just for consumption but for work, storage, and processing consumer waste. Environmental designers are reconsidering conservation and productivity, weighing the implications of conventional aesthetics about waste and cleanliness for the environment and our quality of life, and reintegrating "waste places" into our private and public landscapes. Rich and healthy landscapes are those that include waste places and waste functions as integral parts of their makeup.

Fig. 1.1. *(top)* Waste storage, Metacity / Datatown, MVRDV, 1999. Used by permission of MVRDV, Rotterdam

Fig. 1.2. *(center)* Proposal for a recycling structure in Durant Minipark, Walter Hood, 1997. Used by permission of Walter Hood

Fig. 1.3. *(bottom)* Garbage transfer station Elhorst/Vloedbelt, Almeloseweg, Zenderen, Netherlands, 1995. Client: Regio Twente. Architect: Kas Oosterhuis, Oosterhuis.nl. Design team: Kas Oosterhuis, Ilona Lénárd, Leo Donkersloot, Niek van Vliet, Menno Rubbens. Used by permission of Kas Oosterhuis

Fig. 1.4. *(top)* Sanitation in the Middle Ages. Used by permission of The Fotomas Index, West Wickham, Kent, U.K.

Fig. 1.5. *(bottom)* "New York City—How the Metropolis Invites Disease and Epidemics." From *Frank Leslie's Illustrated Newspaper,* April 23, 1881. Courtesy of General Research Division, The New York Public Library, Astor, Lenox and Tilden Foundations

Fig. 1.6. *(top) The Great Pipe Monument,* Robert Smithson, 1967. Photo by Robert Smithson. James Cohan Gallery, New York. © Estate of Robert Smithson / Licensed by VAGA, New York, N.Y.

Fig. 1.7. *(bottom)* Garbage incinerator, Cheung Chau Island, Hong Kong, 1977. Photo by Catherine Lynch. Used by permission of Catherine Lynch

Fig. 2.1. "The Plumber Protects the Health of the Nation," poster advertisement for American Standard. Photo by Margaret Morgan

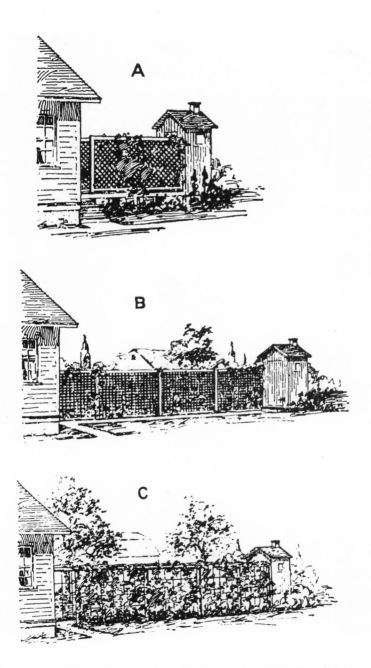

Fig. 2.2. Various tactics to hide and enhance the approach to the privy, such as latticework panels and an arbor, were used in refined, middle-class backyards. Previously published in Ronald Barlow, *The Vanishing American Outhouse* (El Cajon, Calif.: Windmill Publishing, 1989)

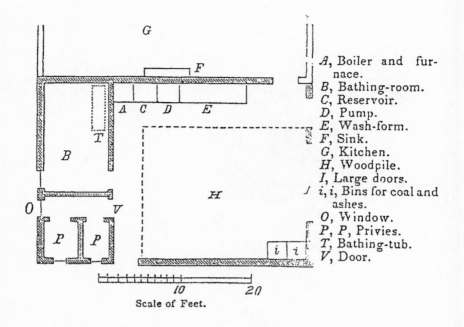

A, Boiler and fur-
   nace.
B, Bathing-room.
C, Reservoir.
D, Pump.
E, Wash-form.
F, Sink.
G, Kitchen.
H, Woodpile.
I, Large doors.
i, i, Bins for coal and
   ashes.
O, Window.
P, P, Privies.
T, Bathing-tub.
V, Door.

```
|.|.|.|.|.|.|.|.|.|.|.|.|.|.|.|
              10          20
```
Scale of Feet.

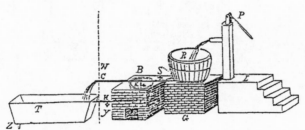

P, Pump. L, Steps to use when pumping. R, Reservoir. G,
Brickwork to raise the Reservoir. B, A large Boiler. F, Furnace,
beneath the Boiler. C, Conductor of cold water. H, Conductor of
hot water. K, Cock for letting cold water into the Boiler. S, Pipe
to conduct cold water to a cock over the kitchen sink. T, Bathing-
tub, which receives cold water from the Conductor, C, and hot water
from the Conductor, H. W, Partition separating the Bathing-room
from the Wash-room. Y, Cock to draw off hot water. Z, Plug to
let off the water from the Bathing-tub into a drain.

Fig. 2.3. Illustration showing an enclosed back accommodation yard (above) containing privies, a bathing space, wood pile area, a water pump, bins for coal and
ashes, and equipment to draw and warm water (below). From Catharine Beecher,
A Treatise on Domestic Economy (Boston: Webb, 1843)

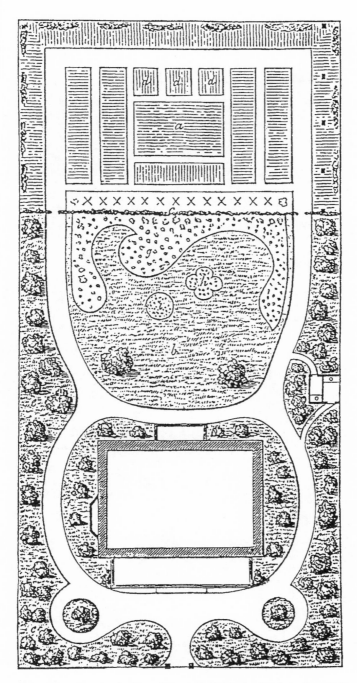

Fig. 2.4. This midcentury yard of a well-to-do individual exhibits a clear division between the large, curvilinear, ornamental garden around the house and the geometric, utilitarian vegetable garden at the back. Two privy compartments are tucked to the side and masked with thick planting. From Andrew Jackson Downing, *Cottage Residences, Rural Architecture and Landscape Gardening* (1842; reprint, Watkins Glen, N.Y.: Library of Victorian Culture, 1967)

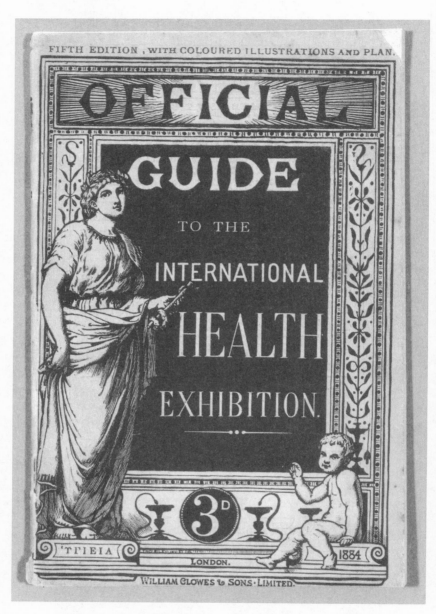

Fig. 2.5. Cover of *Official Guide to the International Health Exhibition* (London: Clowes, 1884). Used by permission of Centre Canadien d'Architecture/Canadian Centre for Architecture, Montréal

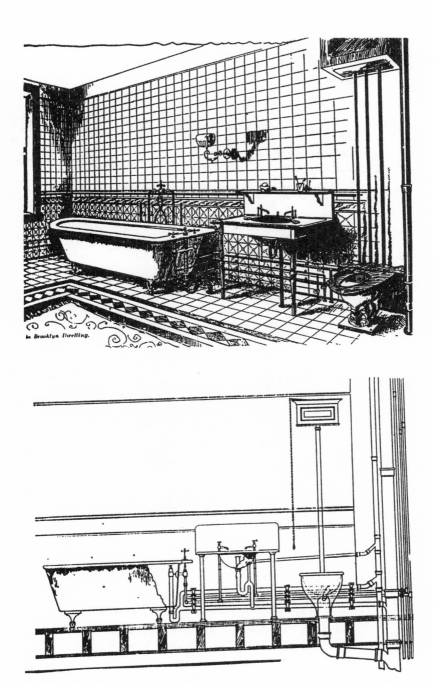

Fig. 2.6. A model Victorian bathroom for a Brooklyn dwelling demonstrates plumbing inventions. Fixtures are raised off the floor and pipes are exposed for the ease of cleaning and fixing. From *Carpentry and Building 7* (New York: David Williams, 1890)

ALL HAIL!

THE SECOND ANNUAL

CLEAN-UP WEEK

For A SPICK and SPAN PHILADELPHIA

APRIL 20th to 25th, 1914

"Why don't they keep the streets a little cleaner?"
You ask with deep annoyance not undue,
"Why don't they keep the parks a little greener?"
Did you ever stop to think that **THEY** means **YOU**?

Fig. 2.7. Environmental reformers in the Progressive era in a clean-up campaign.
From *American City,* April 1915

Fig. 2.8. Back alleys, such as this one, in New York's First Ward, were considered "a perpetual fever nest." From *Report of the Council of Hygiene and Public Health of the Citizen's Association of New York upon the Sanitary Conditions of the City,* 2d ed. (New York: D. Appleton, 1866). Courtesy Avery Architectural and Fine Arts Library, Columbia University in the City of New York

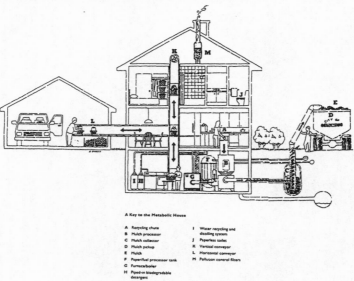

A Key to the Metabolic House

A Recycling chute
B Mulch processor
C Mulch collector
D Mulch pickup
E Mulch
F Paper/fuel processor tank
G Furnace/boiler
H Piped-in biodegradable
   detergent

I Water recycling and
   distilling system
J Paperless toilet
K Vertical conveyor
L Horizontal conveyor
M Pollution control filters

Fig. 2.9. *(top)* An ad for an Adhesive Sealex Linoleum from 1936. "Everything that makes this modern kitchen beautiful makes it easy to clean too: built-in equipment without legs to clean around, smooth and washable walls and counter tops, and sanitary floors of adhesive linoleum." Previously published in Ellen Lupton and Abbott Miller, *The Kitchen, the Bathroom, and the Aesthetics of Waste* (New York: Princeton University Press, 1992)

Fig. 2.10. *(bottom)* *The Metabolic House,* by William Stumpf. A 1989 model for a house with an articulated "metabolic" system shows a kitchen connected with chutes and conveyors for sorting and processing recyclable and mulchable material. Used by permission of William Stumpf

Fig. 3.1. *(top)* Digging up waste samples in Fresh Kills, The Garbage Project, 1989. Used by permission of The Garbage Project, University of Arizona, Phoenix

Fig. 3.2. *(center)* A highly engineered sanitary landfill in San Marcos, Calif. Photo by Mira Engler

Fig. 3.3. *(bottom)* Olympia Park, built on a vast area of a rubble dump for the 1972 Olympic Games, Munich, Germany. Architect: Gunter Grizmek. Photo by Mira Engler

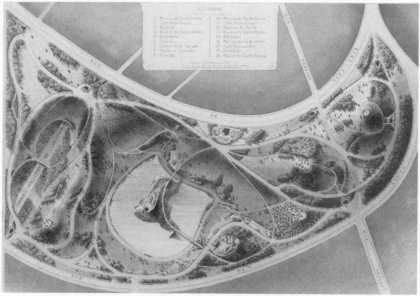

Fig. 3.4. *(top) Escarpment,* Frans de Wit, a mound made of demolition material piled up on the flat polder terrain in Spaarnewoude, Netherlands. Photo by Mira Engler

Fig. 3.5. *(bottom)* Parc des Buttes Chaumont, Adolphe Alphand, Paris, 1864–67, the earliest, most dramatic example of waste disposal site recreated into a park. From Adolphe Alphand, *Les Promedanes de Paris* (Paris: J. Rothschild, 1867–73)

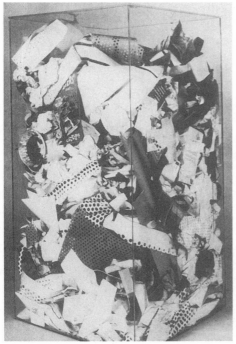

Fig. 3.6. *(top)* A bird's-eye view rendering of post New York Fair, 1964. Used by permission of MTA Bridges and Tunnels Special Archive

Fig. 3.7. *(bottom)* *Lichtenstein's Trash Can,* Arman, 1970–71. © 2002 Artists Rights Society (ARS), New York / ADAGP, Paris

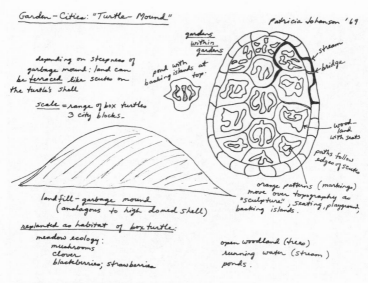

Garden-Cities: "Turtle-Mound"   Patricia Johanson '69

gardens
within
gardens

depending on steepness of
garbage mound: land can
be terraced like scutes on
the turtle's shell

pond with
basking islands at
top.

scale = range of box turtles
3 city blocks.

stream
bridge

wood-
land
with seats

paths follow
edges of scute

land fill - garbage mound
(analogous to high domed shell)

orange patterns (markings)
move over topography as
"sculpture", seating, playground,
basking islands.

replanted as habitat of box turtle:

meadow ecology:
    mushrooms
    clover
    blackberries; strawberries

open woodland (trees)
running water (stream)
ponds.

Fig. 3.8. *(top)* "Garden Cities: Turtle Mound," Patricia Johanson, a 1969 proposal for *House and Garden* magazine. Used by permission of Patricia Johanson

Fig. 3.9. *(center) The Social Mirror,* Mierle Laderman Ukeles, New York, 1983. Courtesy Ronald Feldman Fine Arts, New York

Fig. 3.10. *(bottom) Green Cathedral,* Marinus Boezem, Almere, Netherlands, 1987–96. Photo by Gert Schutte Fotografie. Used by permission of Marinus Boezem

Fig. 3.11. *(top)* *Revival Field,* Mel Chin, Pig's Eye Landfill, St. Paul, Minn., summer 1991. View during first harvest. Sponsored by Walker Art Center and the Science Museum of Minnesota, in cooperation with the Minnesota Pollution Control Agency. © Mel Chin

Fig. 3.12. *(bottom)* Aerial view of Fresh Kills, Mierle Laderman Ukeles, Staten Island, N.Y., 1992. Photo by New York Department of Sanitation. Courtesy Ronald Feldman Fine Arts, New York

Fig. 3.13. *(top)* Entry to Garbage Island, Dekorte Park, Meadowlands, N.J., 1998. Photo by Mira Engler

Fig. 3.14. *(bottom)* *Turnaround/Surround,* Mierle Laderman Ukeles, Danehy Park, Cambridge, Mass., 1993–94. View of the glassphalt path. Photo by Cymie Payne. Courtesy Ronald Feldman Fine Arts, New York

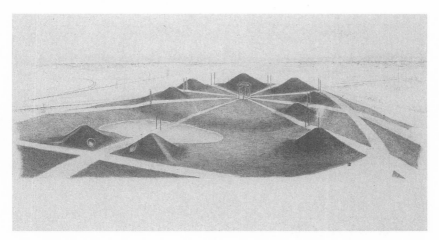

Fig. 3.15. *(top)* Bird's-eye view rendering of *Sky Mound,* Nancy Holt, N.J., 1985.
© Nancy Holt / Licensed by VAGA, New York, N.Y.

Fig. 3.16. *(bottom)* Byxbee Park, Peter Richards, Michael Oppenheimer, and George Hargreaves and Associates, Palo Alto, Calif., 1995. A view from the pole field toward the bay. Photo by Mira Engler

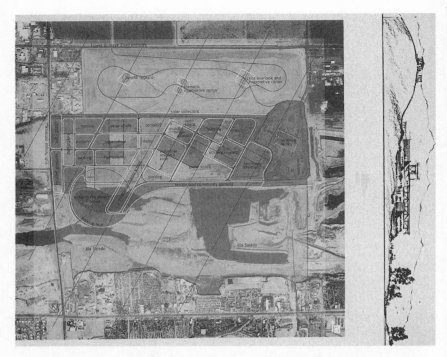

Fig. 3.17. The Center for Environmental Learning and Enterprise. A schematic plan of and section, including the 27th Avenue Landfill. Artists: Lennea Glatt and Michael Singer. Landscape architects: Fritz Steiner and Laurel McSherry. Client: Phoenix Arts Commission and Phoenix Public Works Department. Drawing: Laurel McSherry. Used by permission of City of Phoenix Arts Commission

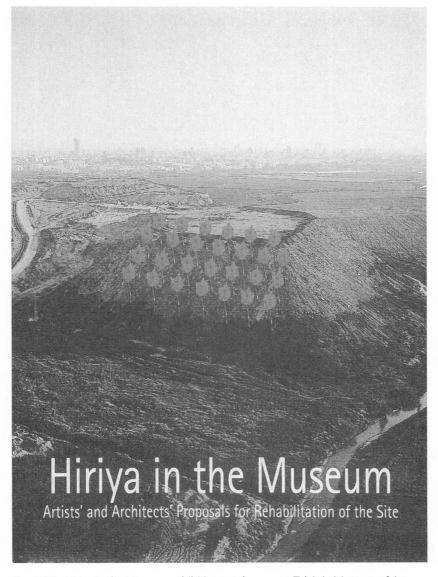

Fig. 3.18. *Hiriya in the Museum,* exhibition catalog cover, Tel Aviv Museum of Art and Beracha Foundation, 1999. Photo by Albatross Aerial Photography, Tel Aviv, Israel. Used by permission of Albatross Aerial Photography

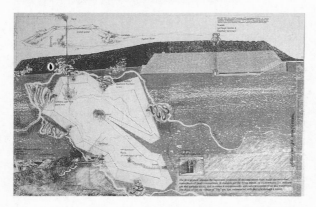

Fig. 3.19. *(top)* "Untitled," Aernout Mik, 1999, contribution to *Tales of the Tip,* Bavel Landfill, Breda, Netherlands. Photo by Rolf ter Veer. Credit: Fundament Foundation / SKOR

Fig. 3.20. *(center)* "Garbage City," Acconci Studio, project for Hiriya garbage dump, Tel Aviv, Israel, 1999. Used by permission of Acconci Studio

Fig. 3.21. *(bottom)* "Mausoleum of Entropy," from "Reclaiming Metaphors Out of the Dump, or Four Gestures for Hiriya," Mira Engler, 1999. Text on drawing reads: "The *first gesture* releases the captivated processes of decomposition from inside the entombed mausoleum of dead commodities. It channels out the dying matter, or its excretions (i.e. methane gas and garbage juice), and re-unites it conspicuously with natural systems of air and waterways. The ruins of the old village of Hir' are, too, re-connected with daylight through a tunnel." Used by permission of Mira Engler

CHAPTER 3

Dumps
Centers of Waste Treatment, Research, and Education

dew shatters into rivulets on crunched cellophane
as the newly-started bulldozer jars a furrow

off the mesa, smoothing and packing down:
flattening, the way combers break flat into

speed up the strand: unpleasant food strings down
the slopes and rats' hard tails whirl whacking

trash: I don't know anything much about garbage
dumps: I mean, I've never climbed one: I

don't know about the smells: do masks mask
scent: or is there a deodorizing mask:

> A. R. Ammons, *Garbage* (1993)

To comprehend the full meaning of dumps requires an understanding of their relationship to the value-laden human concepts and imagination that produce them.

The Many Guises of Dumps
The fantastic dump

Possessed by the fantasy and dread of garbage dumps, the poet A. R. Ammons mines the garbage mound and twines dead commodities into lyrics or, conversely, mines his head and sculpts a garbage mound out of soiled language. In *Garbage* (1993), Ammons metamorphoses what is ubiquitous and repellent into something transformative and meaningful, a cultural mirror for self-introspection. He dedicates the book "*to the bacteria, tumblebugs, scavengers, wordsmiths—the transfigurers, restorers.*"

Few have written so passionately about their perverse reveries of dumps. Garbage is something either to forget or to bitch about. Most people abhor and repress thoughts about garbage dumps. Even those who are attracted to them, embarrassed by their own liking, hide their

stories. Wallace Stegner and Robert Sullivan are among the few who indulge in and convey their ventures into the grime of dumps. In his story
"The Dump Ground," Stegner describes the local town dump as a source
of knowledge where he could read the town's daily habits and history:
"If the history of Whitemud was not exactly written, it was at least
hinted, in the dump. I think I had a pretty sound notion even at eight
or nine of how significant was that first institution of our forming Canadian civilization. . . . The place fascinated us, as it should have. For this
was the kitchen midden of all the civilization we knew. It gave us the
most tantalizing glimpses into our neighbors' lives and our own; *it provided an aesthetic distance from which to know ourselves*" (1962, 43; emphasis added).

Sullivan shares Stegner's fascination with excursions into urban
dumps. In *The Meadowlands: Adventures at the Edge of the City,* he describes
the marvels and uncanniness of New York's longtime dumping ground
in northern New Jersey: "I like to think of the Meadowlands as an un-
designed national park. . . . There were flames everywhere and it was like
Yellowstone with geysers of steam and smoke" (1998, 57, 151). These
once thriving swamps just five miles from lower Manhattan, on the western side of the Hudson River, have become a toxic slough, a receptacle
for urban outcasts marked by human flaws, accidents, and greed. Armed
with a shovel, Sullivan sets out to excavate the dumping ground, to re-
solve some mysteries, and to rescue a buried fragment or two of New
York's great neoclassical Pennsylvania Station and of London's World
War II rubble for his private collection. He marvels at the thought that
some mounds in Kearny Meadows are "London hills," made of buildings
that once stood in the British capital before being transported to the
great American sink of the Meadowlands, as he gropes for the fantastic
and the eerie.

The abject dump

The mystique, eeriness, and beauty of the Meadowlands—or of any
other dump—are necessarily tied to its inexplicable, threatening, and
poisonous nature. Sullivan describes an ambience of despair, death, and
abnormality—lifeless streams of industrial oil slicks, spiritless towns,
rusting auto bodies. He considers the negligence and exploitation that
sucked the life out of the Meadowlands and that bred a deadly legacy,
the plague of twentieth-century industrial society.

Refuse dumps have been victimized and turned into places of abuse and oppression, hosting both victims and perpetrators. The dump is a world where lawlessness prevails and where villains conduct their despicable transactions. It is a place where evidence of unsolved crimes is often hidden, where some human bodies find a resting place. It is where noxious industries and institutions for deviants are found. Snake Hill in the Meadowlands, for example, is the former site of a prison, an asylum, and a sanitarium where tuberculosis patients were quarantined. The outcasts of modern urban life—the immigrants, poor, diseased, and abnormal—who are driven to the edges of the city, are left there to sustain themselves in the midst of havoc. In many metropolitan cities of third world countries in Asia, Latin America, and Africa, thousands of people live, eat, and work in dumps, where they can sometimes make enough to stay above the poverty line. Some eighty thousand squatters cluster around Manila's garbage mountain. In July 2000, hundreds of them died there when the seven-story-high mound, loosened by monsoon rains, collapsed on their shantytown. But the shanties of Promised Land, as the Manila dump is called, are precious homes to their owners, decorated with flower pots and posters of favorite soccer teams (Mydans 2000).

The piteous life of these people is the subject of Latife Tekin's novel *Berji Kristin: Tales from the Garbage Hills.* Tekin poignantly depicts the lives of the poor in the shanties of an Istanbul dump in 1960s Turkey. The rubbish heaps, toxic factories, and deadly winter winds inflict terror, poison, and plagues on the shanties' inhabitants and deny them any dignified form of life. But their human spirit and superstitions, rituals, and poetry are kept alive and sustain their dreams. In those dreams, the earth forms a crust over the garbage hills, new huts rise on the garbage, and flowers of all colors spring up around the huts. Even inside the huts green grass sprouts. They dream of crowds that will then come to Flower Hill to marvel at the colorful blossom and iridescent, gleaming bits of tin and glass (Tekin 1993, 40).

The sociality of dumps

Along with poets, explorers, and the lowly and mean-spirited, dumps have lured and nurtured other kinds of people—speculators, dreamers, and revolutionaries. The risk-taking hopefuls who see prospective gain in future land deals have always targeted landfills and their surroundings as a potential source of lucrative real estate. Their loose surveillance and isolation made dumping grounds hideouts for political escapees and fertile places for new ideas and movements.

Dumps are also places for normal folks who are attracted to their peculiar nature and offerings. Kevin Lynch describes rural dumps as grounds for enacting both disposal and acquisition, places of both social and economic ritual. "Dumping" is the word applied to exploring local dumps for a usable article. Until very recently, writes Lynch in *Wasting Away*, the rural dump was a playful and practical place without stigma (1990, 60, 192). The Sunday afternoon excursion to drop off the family's garbage and perhaps pick up some gossip and a discarded item or two was a well-established ritual. Well-known games such as rat shooting and forbidden love affairs also took place at rural dumps. But the sanitary landfill—cleaner, regulated, and safer—put an end to these and other dump allures. It keeps its trophies out of reach. High fences and soil cover keep out escapees and both gulls and "human gulls" (an idiom coined by Nantucketers).

For sixty years, the town dump on Nantucket, an island off the coast of Massachusetts, served as a disposal and reuse site, trading post, and community center. This social hub, operated by the Department of Public Works, finally ran afoul of state regulations and closed in 1997, to the dismay of many islanders who frequented the place for new possessions. The dump's caretaker, who oversaw the orderly transactions there, was replaced by Waste Options Inc., which capped the unlawful dump and is now running an industrial-type material recovery facility (MRF) at the site, which collects and processes all recyclables and materials. The change brought the demise of the popular rat-shooting ritual but did not restrain the scavenger spirit so deeply ingrained in the island's culture. Hence, the Madaket Mall was inaugurated in 1998 as a citizen recycling drop-off center, where people bring or take clothes and furniture, near the closed dump. Occasionally Waste Options adds its share of up-for-grabs metal and wood piles. On weekend the place is a bustling hub. Boy Scouts collect cans and glass for refund, and regulars keep coming for bargains, keeping the town's tradition and gossip alive in the fitting spot (Crossen 1990; Stevens 1998).

The scholarship of dumps

Contemporary archaeologists have been spared from the ban that puts sanitary fills off-limits. Garbology, or the study of society's garbage, allows academics to transgress cultural codes and trespass the dump fence. The belief that garbage, found as well as missing, eloquently and truthfully describes important aspects of our routine throwaway, have underlain the study of rejected material culture. Like the archaeology of the past, the "archaeology" of the present relies on the cast-offs of culture for its material facts. William Rathje, the renowned garbage archaeologist, has assiduously devoted himself to studying consumer culture's remains, believing that garbage holds a key to the present. His lowly work brought him fame and placed his and Cullen Murphy's 1992 book, *Rubbish: The Archaeology of Garbage,* on the *New York Times* best-seller list. Since 1987, Rathje has dug up garbage from landfills and trash cans to recover its material reality (fig. 3.1) Traditional archaeologists, too, have always looked for the dumps of the settlements, where concentrated evidence of culture can be found. The difference between digging in ancient dumps and present dumps, according to Rathje, is clearly in sensation; those of the past are devoid of the greasy juice and saturated aroma of fresh garbage (Rathje and Murphy 1992, 10). But the future is "more promising," as present landfill engineers labor to keep our buried garbage dry, thereby preserving these sensations for future archaeologists.

Academics digging in dumps must hope to accomplish something beyond indulging their intellectual curiosity. The goal of Rathje and Murphy is effective disposal. They labor to defuse myths about garbage and landfills and set them naked before the reader as physical facts. They want the prevailing fantasies and shortsightedness about waste to give way to data. They claim that this common perception "hampers effective disposal of garbage and leads to exaggerated fears of a garbage crisis" (1992, 84). But can a dump remain a mere fact? Isn't it destined to retain its mystique? Rathje and Murphy dismiss the dumps' cultural baggage, philosophical load, social attributes and allure, perverse rituals, and poetic seductiveness. Unlike Stegner, Lynch, and Ammons, they care little about the aesthetic, social, and symbolic potential of these places.

The ecology of dumps

Archaeologists who labor to recover the culture behind its remains found a small charm in the form of a dung beetle in an ancient Egyptian village. The scarab, a large black beetle that lives and lays its eggs in the dung, was a symbol of fertility and resurrection for the ancient Egyptians. Indeed, dumps are not inert artifacts, devoid of life; they are changing and living environments where the dung beetle and many other organisms make their home. The inner being of dumps is a constant whirl of activity, as microorganisms munch away at the garbage.

Bacteria thriving in dark, underground communities produce temperatures of 100–120° F. They first emit carbon dioxide, then warm, moist methane that feeds fires. In scientific language, an anaerobic process converts decomposable material into humus, methane, ammonia, and hydrogen sulfide. When these products mix with percolating rainwater, the landfill discharges "garbage juice"—more formally, a leachate. As material and energy are exhausted, the dump subsides, taking more than a century to finally come to rest. The fires that result from internal decay keep the surface barren and burned, causing erosion.

Dumps are breeding grounds for rodents and mosquitoes and lure various animals, including wild dogs, bears, and gulls, depending on their location. Rats, mice, and mosquitoes are firmly lodged in the cultural dump of the unconscious. These animals have always been despised and chased away. Other animals are more fascinating. Clouds of gulls swarming over the fresh meals present a powerful scene. The Meadowlands and Fresh Kills dumps, surrounded by remnant swamps, have always attracted an impressive variety of fowl. Robert Sullivan spotted a crowned night-heron and a pied-billed grebe, as well as eighteen species of ladybugs, on one excursion to the Meadowlands. The contemporary human-made wilderness also attracts wild mammals. Wild dogs often seize territory at the dump. When Hiriya, a garbage mountain near Tel Aviv, was still in operation, wild dogs made themselves its guardians. Standing on a small heap by the ascending road to the mountain they growled at workers, trucks, and casual visitors.

Bears make even better stories and an unforgettable spectacle. A dump in Churchill, Manitoba, is known for attracting polar bears that relish rummaging through the garbage and peering into people's living rooms. An attempt to relocate the life-threatening bears failed, as they traced their way 150 miles back to town. The town finally decided to turn the problem into a tourist attraction. Residents now enjoy the dollars left every fall by visitors who flood the town to watch the bears (Lynch 1990, 112).

The open dump is an ecological habitat and a site of accelerated entropy and decomposition. The tamed modern landfill, however, is less a habitat than an artifact, the largest built monument of contemporary society.

The anatomy of landfills

Sanitary landfills are highly engineered places with an impressive infrastructure (fig. 3.2). Standards of venting and draining and a long-term monitoring system shape the landfill structure and cost owners millions of dollars. While preindustrial dumps are generally benign, those of the twentieth century are less innocent, exuding toxic tears and belching gases. Modern landfills are lined with a stratum of impermeable clay under a thick polyethylene membrane that prevents the percolation of liquids. They are crisscrossed with pipes that collect and convey leachate— the toxic broth of chemicals that seeps into surrounding soils and renders groundwater and rivers poisonous—to a place where it can receive decent treatment. Some landfills treat their own "juice" with a small conventional facility or a "natural kidney"—a marsh—though many direct their leachate to the city sewage treatment system. A network of long, vertical, perforated pipes extracts methane gas and then burns it as it is released to the air, or channels it to produce energy. The gas wellheads are extended as the landfill rises, poking through the surface of the rolling hill and looping back downward to a collection system; gas vents in landfills without a gas collection system often look like a grid of periscopes jutting above the water. Unless vented and collected properly, methane can migrate through fissures and cause fires and explosions.

Landfills are restless, powerful, and orderly operations whose engineers carefully build "cells" and "stories." Like swarming bees, every day hundreds of daily trash and dirt trucks and tractors haul, dump, shovel, and compact the garbage, covering it with a layer of dirt. The daily garbage typically is coalesced into a cell about 30 feet wide by 25 feet high by 100 feet long. To disperse rainfall efficiently, layers are built to a preplanned height and slope, and contours are shaped according to strict specifications. Planners assume that the drier the landfill is, the less risk it poses to groundwater. A cap, made of a thick layer of clay that sometimes covers another sheet of polyethylene, makes a perfect seal for the dump. Finally, a thinner growing medium of fertile soil with grasses for erosion control is placed on top, like the icing on a cake.

Even the builders cannot solve the landfills' mysteries—for example, why capped landfills are littered here and there with car tires. Though once buried deep, the tires are somehow ejected to the surface. Scientists are still hypothesizing about this enigma.

The narratives of dumps

So far we have mined the dump for its poetry, fantasies, viciousness, sociology, ecology, and anatomy. Others mine the dump for narratives. In *Landscape Narratives* (1998), Matt Pottieger and Jamie Purinton consider a dump (and other wastelands) as a site caught in a tragic plot of the disrupted natural course of events. As a wasteland, it evokes five interrelated narratives, which they name death/elegy, ecological history, progressivism, crime story, and entropy.

The first two narratives are distressing and sorrowful. The elegy tells the story of the apparent loss of once-living species and habitats that inevitably accompanies wastelands. It yields mourning and grief. The natural history of the site—for example, the evolution of the Meadowlands from the original marsh and cedar swamp habitat to an area dominated by foreign, invasive, and inferior reeds—hinges on an ecological trajectory of time and change. These two narratives readily serve the guilt-laden altruists of nature who seek to return such sites to their prior ecology. The third is a counternarrative of optimism, progressivism, and victory. It describes the impulse to turn a wasteland into valuable real estate and put (retired) dumps to some kind of use. Ironically, dumps were often used to "improve" marshes by making them useful spaces, which in turn became wastelands. The last two narratives are again tragic and hopeless. The crime story of wastelands (akin to the abject dump) speaks of geographic marginality and profanity, of places that nurture criminal transactions and negligence alike—illegal dumping, pollution, unsolved mysteries. Finally, the entropy narrative relates to chaos and loss of energy, as accelerated processes of change at the dump exhibit mutability and messiness.

Pottieger and Purinton claim that these associations with death, oppression of nature, heroic development, corruption, and loss of order are unconscious and exercise great power over us. They condition our attitudes toward garbage and shape the design and politics of dumps.

Dumps are fraught with ample, conflicting imaginings and facts; they yield scientific data and are laden with myths; they are built as monuments to last and host processes of entropy; they symbolize death while standing for progress. Their power lies in their contradictions and ambiguity. Probes of the dumps reveal treasures, fears, losses, abnormalities, absences, and enigmas. Their inherent dialectics makes them a rich subject for intellectual and aesthetic investigation of our everyday lives and places. But what is the place of dumps in relation to other places we have created?

The Meaning of Dumps

Ideology of the "dump"

Dumps are the epitome of marginal places. Their difference—poisoned, entropic, and unpredictable—provides them with the distance to serve in the cogent role of the Other.

The pioneering work of Robert Smithson demonstrates this very idea. Smithson sought to explore and illuminate the dialectical potential of marginal landscapes and their potent role as sites that permit the deconstruction and reconstruction of relationships between culture and nature (see chapter 1). Although Smithson did not directly engage dumps (except in *Heap of Language*, 1966), his projects nevertheless are set in damaged landscapes, suggesting his subversive approach to maligned places.

Michel Foucault's discussion on Heterotopic places (see chapter 1) also does not refer to dumps; however, based on the characteristics he defined, they are well fitted to this category of Other Spaces, acting much like Heterotopias in the roles they play in the matrix of our environment and lives. Heterotopic places, such as cemeteries, prisons, fairs, brothels, and museums, exist in every culture and function in ways that are clearly defined but can change over time. They lie outside the time and space of other places. The boundary of each is well demarcated; though it is penetrable, those who enter do so either by force (e.g., into a prison or psychiatric hospital) or by first completing rituals (e.g., of purification, before entering a cemetery). Heterotopias also lie outside real time. They are either immune to the wear and tear of time (e.g., a museum) or they suspend time by focusing on futile, transitory, or precarious nature (e.g., a fair) (Foucault 1997).

The cemetery and the dump are alike. Both are part of our disposal system. Both are connected to all people, the ultimate resting-places for them and their belongings (except for items that find their way to the museum). As strange as it may seem, the geographic history of cemeteries and dumps also coincides. Both were first located at the yard—church or home—and then forced to the town's periphery. Cemeteries were adopted as the first urban parks in the United States, and many landfills are now turned into parks as well (as at the cemetery, the lawn eventually covers the content below). The dump or landfill also resembles the museum, storing artifacts in its time capsule; the landfill accepts and entombs our unwanted vestiges, while the museum encases what has been determined precious. A prison keeps deviants away from society; a dump locks up the unwanted and the rejected.

Heterotopias, dumps included, are created to contain things and people we set aside, whether despised or cherished: the dead, the deviant, and the outcast, or the prized and valuable. Heterotopias affirm and sustain ordinary places—accepted behavior, the useful, the normal, and the desired—but at the same time, they expose them naked and disclose their faults and ambiguities (Foucault 1997, 356).

"Dumpy" etymologies

The etymology and meaning of the terms in use offer yet other insights into dumps. "Dump" has been lost from the genteel lexicon; its terse and thudding sound is rarely heard. But the word remains utterly direct and expressive in rendering the essence and tone of our great dumping grounds. The euphemistic "sanitary landfill" presents itself as a completely different creature while its function stays essentially the same.

Dump has always connoted a depressed, low, or dull condition. It was first used in the sixteenth century, seemingly derived from the Dutch *domp* (to become or to be at a dazed or puzzled state of mind") and the German *dump* ("a dulled, melancholy or depressed sensation") (*OED*, 1989). By the late eighteenth century, the term was applied to an abyss or a deep hole in the bed of a river or pond and to various objects of "dumpy shape." Dumpy objects were rough leaden counters, or coins, or metal blanks before they are coined. Whether object or hole, a dump was a low and dull entity. Phrases such as "not care a dump" meant to not care at all, signifying unimportance. In the United States, miners in the West began calling their waste piles "dumps" in the nineteenth century, and the meaning later expanded to include any place where refuse from a mine or quarry was disposed.[1] It finally became a pejorative or contemptuous term for any place or building that is neglected and trashed. To "take a dump" and "What a dump!" are idioms that almost every American is familiar with. The latter is attributed to actress Bette Davis (in *Beyond the Forest*, 1949); the clip has been played and parodied often.

Any vacant lot or roadside has the potential of becoming an informal dump, absent a No Dumping sign specifying some penalty. The formal or informal garbage dump, or open dump, is a natural land cavity or pit where material is simply piled up to decay. Dumps, considered a major nuisance and source of epidemics, became the target of massive eradication efforts in the late nineteenth century. But even as late as the 1970s thousands of instances of unauthorized land dumping were still taking place. Early in the twentieth century, experiments using garbage for "landfill" began. These were the first "land reclamation" projects to use garbage to "conquer" bays, rivers, and marshes (Melosi 1981, 168–69; 2000, 269). It was believed that landfills could make useful ground out of low-lying areas, gorges, gullies, wetlands, and particularly swamps known for their disease-ridden air. Schemes to fill out human-made "land holes," such as abandoned strip mines and quarries, were considered and in some cases put into practice.

Sanitary fill (or landfill) started in the 1920s as a controlled dumping method: garbage was layered with compacted earth mixed with street sweepings, and ashes were laid on top. It originated in Great Britain, where it was called "controlled tipping." (The word "tipping" recently became a professional term for dumping.) When Americans devised their version in the 1930s, they called the result "sanitary landfills" (Melosi. 1981, 219). The first sanitary landfill in America was built in Fresno, California. In 1937 Jean Vincenz, the public works director of the city, put to work the trench method, in which garbage was covered every day by dirt and compacted. This approach to dumping became common after World War II; it remains today the dominant method, in the improved form of the "secure sanitary landfill," which requires stringent standards for groundwater protection and methane collection. Three other terms—heap, tip, and tail (or tailing)—are often used as synonyms for dump.[2]

In the 1960s, various rising mountains of garbage deservedly received the title "Mount Trashmore." The label was first applied in 1965 by the author John Sheaffer, who used it to refer to a garbage mound in the Blackwell forest preserve in DuPage County near Chicago (Johnson 1969). Katie Kelly, author of the captivating book *Garbage* (1973), cites another pioneering example, a 1967 towering garbage mound in Virginia Beach, Virginia. The two are rival contenders for the title of the "original Trashmore." Unlike dumps in low terrain, sinks, or swamps, garbage hills cannot escape public attention. They are relatively recent additions to the landscape, and they require an aesthetic resolution and special plans for their afterlife. The Blackwell mound was designed to be part of a recreational park, which also included a nearby lake and toboggan runs. The "vertical" landfill was lined with a thick layer of soil that had been excavated in building the lake. County residents considered both lake and mountain to be wonders of the world. Kelly tells us that the 161-acre mound in Virginia Beach was planned to host an amphitheater, ball fields, and a parking facility. Proud citizens donated labor and their own garbage to construct it (Kelly 1973, 94). Today it boasts a park with two lakes and a 68-foot high observation point; in 2000, it accommodated 800,000 visitors. Trashmores in Europe began earlier, as heaps shaped from bomb rubble of World War II.

Types of dumps

Garbage dumps contain indistinguishable ensemble of discards that are stripped of their individual identities, cut off from their trajectory of usefulness, and forced together into the mess. When we speak of garbage dumps, we mostly think of those that contain our personal castoffs, the refuse we place daily in the garbage cans—things that are leftover, used and abused, dry and wet, organic and inorganic, toxic and harmless. In professional terms this commingled material is called municipal solid waste (MSW). But some waste goes into more exclusive, specialized dumps.

The logic of specialized dumps relies on the sheer quantity of the material and the ease of managing it. Demolition and industrial waste dumps have existed for quite a while, for demolition material is both abundant and inert in nature; these qualities make it practical to dispose of separately, process, or reuse as aggregate. The nontoxic and dry bulky matter promises a fill without pollution and the hazards and subsidence associated with other landfills, thus making possible multiple options for its end use. In the past, construction waste was used for land reclamation along bays and rivers. San Francisco and surrounding cities gained new lands from the adjoining water, as did New York City. Germany forged a tradition of raising rubble mountains of war debris. Berlin is among the cities adorned with several rubble heaps. Its "Mount Junk" and Munich's Olympia Park are hills sculpted out of war rubble. Mount Junk rises 360 feet into the air; it has grapevines growing on one side, an observation post on its summit, and ski and toboggan runs on its slope. Olympia Park, designed by Gunter Grizmek as the site of the 1972 Olympic Games, is built on a vast area of a rubble dump. Structures bombed into heaps during World War II were reshaped into steep hills to echo and work in concert with the tented silhouettes of stadium structures (fig. 3.3). The mounds were then carpeted with a lawn and planted with a few clumps of low trees in order to accent the hills' spectacular stature, while burying the "drama" beneath.

Escarpment is perhaps the most inspiring park on a rubble dump. The mound, made of demolition material piled up on the flat polder terrain in Spaarnewoude, Holland, was sculpted by the artist Frans de Wit into an evocative monument of bared rubble that conspicuously exposes its content. A narrow and steep ascending stairway is laid between double concrete walls, piercing the mound in the center. The views from the stair flight to the distance are tightly framed and breathtaking. In *Escarpment* the waste itself, retaining the narrative and dialectic of the place, is used as an unmediated, expressive element in the park; it also generates activities such as rock climbing and experiences of discovery (fig. 3.4).

Recent regulations in many localities banning organic materials from municipal solid waste landfills should be credited with creating dumps of yard waste or, more precisely, composting and mulching centers. Oil from car engines, tires, and toxic waste are stored and processed separately. Industrial tailings or slag heaps are another kind of dump. Their dry and poisonous content contains metal compounds that are harmful to the land and groundwater if not handled properly. Europe has an abundant inventory of slag heaps in its vast industrial regions devoted to coal mining and steel; some date back to the eighteenth century. While many slag heaps in America are found in remote locations, Europe's are close to and visible from towns. It is only since the 1990s, though, that their reclamation has been contemplated. Germany initiated several design/art competitions and projects to shape old and new slag heaps in the industrial Ruhr region as earth art or land monuments. The International Building Exposition of Emscher Park (1989–99) considered the artificial hills in the area as opportunities for a new regional landscape identity. In the same spirit, Halde Schurenbach was crowned with a monumental artwork by Richard Serra, and the Mottbruch waste heap in Gladbeck is to be shaped as changing sculptural volumes during the ongoing dumping process (Dooren 1999, 101).

With ever more complex politics and new moratoriums and measures blocking some dumping and requiring in other places that it be environmentally safe, specialized dumps containing and processing sorted wood, metals, glass, and so forth are likely to dominate in the years to come.

The politics of dumps

The politics of garbage and its designated sites has always been complicated. Although garbage collectors (or "sanmen," as they are sometimes
called) have very low professional status, their strikes are acts of great
power, a nightmare for any politician. Public officials have risen and
fallen on their ability to manage waste. Garbage provokes and mobilizes
the public.

Beginning in the late nineteenth century, the key players publicizing
the risks of garbage in American cities were sanitarians and health experts, on the one hand, and reform movements led primarily by women,
on the other. The first group relied on science to emphasize the rational
connection between garbage and health and placed trust in technology;
the second group used aesthetic and moral arguments and considered
the condition of city streets to be an extension of housekeeping. Wom-
en's domestic values of order and cleanliness were effective tools to rally
the people in demanding public street cleaning and better services.

In the 1920s lay people in reform movements were replaced by specialized sanitary engineers. Waste management was taken over by municipalities, as federal incentives and regulations bolstering production
and consumption resulted in increased waste and overwhelmed local
governments. Garbage was politicized in the years following World War
II, as big business began influencing and shaping garbage management.
Later, as the link between garbage and pollution became clear, federal
and state governments intervened, recognizing that waste disposal has
become a problem of national scope.[3] Profit-driven industries interested
in stimulating consumption and expanding manufacture began object-
ing to new regulations aimed at reducing pollution and blaming customers for the environmental problems. Big business slowly began taking control of waste management away from municipalities (Luton
1996). The rhetoric consequently shifted from presenting the city holistically to portraying a nation at war with its garbage (Blumberg and Gottlieb 1989, 5).

The dynamics became even more involved in 1987, when the public,
prompted by the media, was awakened to a garbage crisis. The infamous
unclaimed garbage barge *Mobro* crystallized the politics of garbage disposal.[4] But it was the NIMBY (not in my backyard) syndrome that most
changed the waste management scene and the map of disposal sites. This
grassroots force, which often draws on fears about possible damage to

community safety (because of health risks from toxins), quality of life (because of the traffic nuisance posed by garbage trucks), and, not least, property values, is a social phenomenon of much power, though of ambiguous ideological character. Like big business, it acts in its own interests, not to advance a rational solid waste policy (Lifset 1990, 5).

New laws and higher standards of waste disposal sent the expense of landfill infrastructure, tipping fees, and closure costs all sky-high. This trend further cemented the shift of landfill operation and waste collection from local municipalities and small companies to the hands of a few private multimillion-dollar companies. Solid waste management has become a thriving monopolistic business. Landfills became even further removed physically from most people, increasing the literal (and mental) distance between waste and its main producers. The new regulations also led to a desperate search for accommodating "landlords." Thus, the business of garbage export to third world countries became even more profitable; the use of antiquated and more harmful landfills continued; and areas of the country with less restrictive regulations, less political power (and wealth), and more open space were obvious targets for landfill sitting. The garbage of the strong and wealthy came to rest in the backyards of the weak and poor (Weintraub 1994).[5] Though the money gained by accepting the rejects of others was an essential part of this exchange, in many cases the choice was not simple. Now, almost anywhere in the United States, waste companies must negotiate long and intensely with local officials and residents for the permits they need to operate. Beyond the costly technology necessary to run the landfills, the corporations must pay for mitigation and beautification projects. To that end they harness the conciliatory and transformative skills of environmental artists and designers. Today's new landfills are sold as amenities, as a temporary disturbance that eventually will pay off for the adjacent community.

Though dumps may have always signified the other side of civilization, they were in fact evidence of an urban cultured society.

Terrain for Design Inquiry
Premodern dumps

Formal dumps in antiquity were rare and existed only in the most civilized nations. By 800 B.C.E. organized dumps were found in Jerusalem, which used a valley called Gei-Ben-Hinom (literally, the valley of "hell") outside its walls, and in 500 B.C.E. Greek cities ordered all garbage removed to designated sites at least one mile away. The citizens of Athens hauled out their fecal matter, potsherds, and animal carcasses, as well as unwanted babies, and left them to decay at the dump. Rome had a town pit at its edge and forbade disposing of feces in open carts and ditches. It also was the first to organize workers to scoop the nuisance from the streets (Kelly 1973, 16). Yet places like Paris and London did not continue the example of these great civilizations of the past. Until well into the Middle Ages, people dumped waste of all kinds conveniently where it was generated, near their dwelling or workplace. Before the late eighteenth century, all space was treated as a dump. If today's garbage mounds rise both inside and outside of cities, it is only because cities expand and merge and therefore must embrace and absorb their own previously despised, peripheral grounds. In the distant past, however, cities rose in isolation on rising remains of centuries of garbage and city ruins. Some Middle Eastern cities became massive mounds, called "tells," looming above their surrounding plains. As settlements became crowded, streets became natural receptacles, strewn with garbage and offal and frequented by rats and stray pigs. European cities instituted street cleaning services only in the sixteenth century and formal dumps in the late eighteenth century.

As cities grew dense, so did their garbage. Streets were filthy and rings of refuse mounted around city walls. Trash-filled moats endangered cities' own defenses. One of the earliest signs of concern dates to 1388, when the English Parliament enacted a law forbidding the disposal of garbage and fecal matter into rivers, ditches, and other bodies of water. London forbade its citizens to dump any dung within the confines of the city or to carry it before 9 P.M. (Kelly 1973, 20). But such laws were exceptions rather than the rule, and they were generally ineffective. In the sixteenth century as more cities tired of the stench and sliminess of their roads, they instituted laws prohibiting dumping into the streets and waters. In 1539 Paris issued a decree requiring each home and business owner to sweep in front of his or her building every three hours, starting at 6 A.M., and "heap his refuse against the wall, or place it in a basket or some other receptacle until such time as the garbage collectors make their rounds, or face a fine of 10 sols" (Laporte 2000, 29). Two garbage collectors and a cart were assigned to each neighborhood, and they announced their arrival with a bell. Dominique Laporte frames this act (and other rules concerning human waste) not only as a solution to a problem but also as an indicator of a new age and worldview and the emergence of individualism. Though they signified the beginning of an important change, such laws were hard to enforce and had little impact on the city.

By the eighteenth century, ordinances requiring that dumps be located outside the city became more urgent and prevalent. The sites chosen were low and depressed natural areas and human-made cavities. In the 1780s, Paris was surrounded by twenty-four dumps, many of which manufactured ammonium sulfate out of cesspool contents and continued to foul the air with fumes (Reid 1991, 72). Some were former quarries (e.g., Montfaucon) and adjacent to slaughterhouses. In the early nineteenth century, cities relocated their dumps farther away and paid a greater attention to sanitary conditions and services. Cemeteries and slaughterhouses were moved from the inner city to the outskirts.

The sparsely populated American colonial towns were generally cleaner than their European counterparts.[6] Most had only crude sanitary regulations by the late seventeenth century. There, too, town dwellers cast rubbish and slop into the streets and alleys. Some restrictions on trash disposal practices began to appear during the eighteenth century. For the majority of American cities, collection of household and commercial refuse by scavengers and the individual dumping or burning of garbage, ashes, and rubbish provided adequate sanitary treatment.

Here, too, natural "ready-made" sinks—gullies, gorges, marshes, and swamps—were filled up first. A lake, an ocean, or a river made a good trash receptacle. Then, human-made sinks—quarries and pits—were used. Boston located its open trash and sewage area on swampy Columbia Point at the edge of the city below South Boston. Chicago used Lake Michigan, three miles south of the Chicago River. New York dumped its garbage into the ocean.

The sudden growth and crowding of cities in the mid-nineteenth century exacerbated the waste problem, making it not moderately bothersome but unbearable.

Modern dumps as urban parks

The acute circumstances of the mid-nineteenth-century urban crisis—smoke, noise, crowding, crime, and epidemics—brought putrefying garbage to the center of consideration. The volume of street sweepings, horse manure, ashes, and household rubbish grew significantly and created nightmarish urban scenes. More attention was given to sources of bad smells. Health officials publicized the connection between rotten matter and disease, harnessing civic awareness to support the modern sanitary project. Nevertheless, the dominant disposal methods at the time were still to dump refuse into water and on land, often into vacant lots near the least desirable neighborhoods (Melosi 2000, 18). In some cities waste was also used as a fertilizer or animal feed, and the swill yards and feeding areas on the margins became new sites for breeding nuisances. Cities that could directed their refuse to islands where the processing industries were concentrated. Randalls, Rikers, and Wards Islands were city repositories for New York, Spectacle Island for Boston.

Once municipal officials recognized that the waste problem and its subsequent health hazards affected the quality of city spaces, the efforts of professionals in two seemingly opposite professions—sanitary engineering and landscape architecture—came together to transform dumps into parks. It was believed that parks could introduce the quality that was found in the disease-free country. Dumps were the city's open wounds; the magical cure was to cover them with green parks.

The elegant Tuileries Gardens in Paris were laid out on top of the dump just outside the palace's grounds. Most notable was the Parc des Buttes Chaumont, which was built between 1864 and 1867 over the notorious sewage dump and slaughterhouses in Montfaucon. Like its successors, the dump was viewed as an immoral world, a threat to public order, and a potential source of contagious diseases. But this primary symbol of evil and sin could become its opposite. After eighty years of operation and intense debate, the dump site was closed and transformed into a magnificent park. During Paris's grand projects of infrastructure and parks under the auspices of Baron Haussmann and in preparation for the 1867 World Exhibition, Jean Charles Adolphe Alphand, who studied engineering at the École des Ponts et Chaussées, was charged to design the new park. Alphand's innovative model park erased all traces of garbage and created a perfect leisure ground with dramatic spectacles of lakes, cliffs, and bridges (fig. 3.5). During that time, claims architectural historian Maria Luisa Marceca, the concepts of urban residual space and urban unsanitary, unhealthy conditions were linked, and land residues were viewed as key to the resurgence of the city (1981, 59). Alphand drew an analogy between the processes of the city and human physiology and suggested that human progress and technological progress brought about health. The park was a means to effect this change. The environment was reclaimed biologically; nature entered the city and simultaneously embraced the machine. The design acknowledged technology—railroad and roads—uniting city and nature while revealing the panorama of the city from its bucolic buttes and employing dramatic elements and metaphors (Meyer 1991, 21; Marceca 1981, 58).

At the same time on the American continent another visionary landscape designer was embracing New York's marginal grounds and dumps and bestowing bold, innovative parks on them. Frederick Law Olmsted redeemed the city's northern wasteland of slaughterhouses and other noxious establishments by creating Central Park. With the aid of new technology, he, too, sculpted and moved enormous volumes of dirt, accented rock formations, and established a grand avenue for the horse-drawn carriages that would bring New York's bourgeois to the urban leisure ground and reform the urban masses.

The close link between parks and urban sanitation was an important connection throughout Olmsted's career. During the early 1860s Olmsted served as executive secretary of the United States Sanitary Commission, an agency that was a training ground for many prominent sanitarians (Schultz 1989). In fact, Olmsted, Horace W. S. Cleveland, Robert M. Copeland, and other landscape architects of the time believed that parks in the central city not only enhanced the livability of the city and the moral standards of its citizens but also literally erected barriers against diseases and mental distress. John Rauch, an early Chicago park planner, declared in 1869, "A pleasing landscape . . . exhilarates [man's] spirit, while a dreary waste . . . produces depression" (Schultz 1989, 156).

By the time that Robert Moses came to power as New York City's park commissioner in 1933 he had two great precedents to draw from—Parc des Buttes Chaumont and Central Park. Beginning in the 1920s, Moses dreamed of a second great park for New York (one that would be grander than Central Park and would carry his name). The site he chose was Flushing Meadows, a former marshland that had become a repository of the city's refuse, the Corona Dump, and (except for the name) he largely realized his dream between 1939 and 1964, using two world's fairs as his vehicles (Caro 1974, 1082). In the 1920s the area glowed with fires and was infested with rats. F. Scott Fitzgerald described the place in *The Great Gatsby* (1925) as "a valley of ashes—a fantastic farm where ashes grow like wheat into ridges and hills and grotesque gardens, where ashes take the forms of houses and chimneys and rising smoke, and finally, with transcendent effort, of ash-gray men, who move dimly and already crumbling through the powdery air" (quoted in Caro 1974, 1083). Like

Alphand, Moses hoped to transform the symbol of evil and ugliness into one of progress and beauty.

A small part of the park was opened in 1939; it was expanded and completed for the 1964 World's Fair with much financial struggle and controversy. The park was envisioned as the "Versailles of America" and laid out in a formal, geometric Beaux-Arts plan (attributed to Gilmore D. Clarke). The south section was informal and consisted of two curvilinear, artificial lakes (Miller 1989) (fig. 3.6). The many pavilions and exhibits followed the theme of the fair, "building the world of Tomorrow," focusing on developments in communication, travel, and public health. More than a billion dollars was invested in the one-square-mile Corona Meadows, which spurred the construction of new roads and bridges, supporting a new era of speed and automobile.

In 1947 Moses eyed a tidal wetland of salt and freshwater marshes, an area on Staten Island owned by the city's Parks and Recreation Department, for New York's new landfill site. He imagined that Fresh Kills would be converted into a future amenity for the city—a park and a municipal airport. This heroic urban gesture was carefully fit into a grand vision for the great city, and Moses sold his plan to the borough president by promising to build the West Shore Expressway. He also promised to stop dumping in 1951. The expressway was built, but dumping continued in Fresh Kills until 2001.

In the first half of the twentieth century, many other closed dumps were recycled into salable land by developers; they were covered with parking lots, airports, shopping centers, industrial land, and even low-cost housing units, some of which were later slated for demolition because of structural hazards caused by subsidence of the fill. Other dumps were converted into public parks or golf courses, thereby remaining open spaces. In all cases, the dump was redeemed, the repugnance covered, and the bad memory erased. Landscape architects, the apostles of landscape aesthetics, were typically given the task of masking urban wastelands, and they readily used nature-like elements to wipe out guilt and make dumps more palatable. Modern technology enabled these transformations and put an end to the misery that premodern dumps had inflicted on urbanites. Garbage dumps, renamed sanitary landfills, no longer transmitted germs directly to populations; they now began releasing their poison indirectly and through the environment.

The fulcrum of change

Sanitary landfills began to dominate garbage disposal practices in the late 1940s; the intermittent layers of soil now added to the waste only facilitated social amnesia. As garbage mountains rose and awareness of environmental issues heightened in the 1960s, the garbage problem received renewed attention. The modern agenda and its outcomes were now scrutinized and criticized through the lens of postmodern thought. Artists and designers began pointing to the links between waste problems, failure to perceive waste, and landscape aesthetics. Several artists began to challenge traditional taboos and treat waste disposal sites as settings for artwork, as conspicuous monuments in the landscape. Their work was preceded by anarchist and counterculture art movements, including Fluxus in Europe and pop art in both Europe and America. Claes Oldenburg and Arman, for example, used mundane, mass-produced, and discarded commodities as art objects and as subjects for contemplating and commenting on consumer society. In 1960 Arman filled a Paris gallery with so many heaps of garbage that it was impossible to enter the exhibition. The act exemplified the pseudo-biological cycle of production, consumption, and destruction. He then encased trash in Plexiglas enclosures that he offered as works of art (Silk 1993, 69) (fig. 3.7). Earthwork artists also established a threshold that legitimized artistic experimentation with unsightly, degraded, and dangerous landscapes. In particular, the works of Robert Morris, Robert Smithson, and Alan Sonfist drew attention to abused urban and industrial landscapes and the consequences of contemporary habits of production and consumption. Calling on art to shun elitism and perfection and to deal with decay, pollution, and devastated wasteland, such as quarries and toxic land, they contested the earlier camouflage and restorative approaches.

Two landmark projects signified a new beginning in the history of dump reclamation. In 1962 the urban designer Brenda Colvin offered a plan for Gale Common, a pulverized fuel-ash disposal site near a coal-fired power station in Yorkshire, England. Her poetic response was manifest in the creation of a new landscape feature, an artifact clearly distinguished from the flat surroundings. The steep hill was conceived with modulated, spiraling terraces and geometric patches of crops (Moggridge 1993, 23). In the United States in 1969, artist Patricia Johanson designed *Garden Cities: Turtle Mound,* a proposal to reclaim a garbage dump as a park in the form of a turtle. It included intimate spaces for people and wildlife. Commissioned by the magazine *House and Garden,* the project was never built, but it has influenced later waste-related site design (fig. 3.8). In the 1960s, however, such ideas were the exception.

During the 1970s more American artists engaged ecological art to cope with polluted water and land, but only a few involved themselves directly with solid waste disposal. Most significant was Mierle Laderman Ukeles, who has drawn unparalleled attention to the routine and failing maintenance of our cities. *Manifesto for Maintenance Art* (1969) lay the philosophical foundation for her early works, such as the documentation of the New York City sanitation system and facilities and a proposal to transform six of New York's forgotten dumps into urban earth sites. Ukeles rescued waste sites from obscurity. She saw these leftover, despoiled wastelands as abstract symbols of the authority of the city and recognized in them an inherent conflict of public practice and private needs. For Ukeles the dump is "a rich, awesome zone, highly charged and vibrating, awaiting the entry of art," and art is an articulator, a mediator, a healer, and a creator of a new reality (Ukeles 1992, 12).

When in 1979 she became the unsalaried artist-in-residence with the New York Department of Sanitation, she completed several art installations and performance pieces that shed light on the sanitation system and the people who make it function. In *Touch Sanitation* (1979–80) she walked thousands of curb-miles with sanitation employees and shook the ungloved hands of all 8,500 sanitation department workers over an eleven-month period, thanking them "for keeping New York City alive." The handshakes restored the workers' dignity and brought some of them to tears. In *Garbage Truck Dance* (*Vuilniswagendans,* 1985) she created a ballet performance with ten garbage trucks and four street sweepers in Rotterdam. In *Social Mirror* (1983) she flanked the sides of a garbage truck with mirrors and had it driven by a uniformed sanitation worker in a New York parade, compelling bystanders to see their own image reflecting from the passing truck (fig. 3.9). These works forced direct confrontations between what is perceived to be the lowest of culture and the highest of art and between citizens and their waste services.

Ukeles's longtime romance with the mother of all landfills, New York's notorious Fresh Kills, has generated imaginative, original research into the multivalence and complexity of dumps. Since 1977 she has followed, studied, and shaped a vision for Fresh Kills, working relentlessly to draw public attention to the horrors and treasures of this and other dumps.

For her the landfill is a fascinating hub of activity and a significant cultural milieu that must be stripped of its stigma and given full heed.

As more artists began using the landscape as a legitimate sculptural medium, more landfills became coveted subjects for exploration. In the spirit of the 1970s earthworks genre, the Seattle Art Museum created in 1979 an exhibition and symposium titled *Earthworks: Land Reclamation as Sculpture* and invited seven artists to engage a variety of degraded landscapes, such as quarries and dumps, in King County, Washington. The project accented the effects of human practices on the landscape and cast the artists as manipulators and sculptors of earth. In the 1980s artists began staging their work on landfills, employing them for social critique and environmental action. During the summer of 1982 Agnes Denes planted and harvested a wheat field on top of a New York landfill created by the land excavations and demolition for the construction of the World Trade Center. Using the twin towers as a backdrop, her temporary art *Wheatfield: A Confrontation* (1982) exposed the dilemma of the global financial world and the state of agriculture and land stewardship.[7]

In addition to their inherent symbolism, the conspicuous landfill earthworks engendered an aesthetic debate over their visual-formal qualities. In Europe, as early as 1973, the Dutch periodical PLAN featured a joint workshop of landscape architects and representatives of waste organizations that explored issues of waste and landscape and challenged professionals, as one engineer noted, "to reproach the refuse world for 'making fancy creations with tufts of trees on them'" (Dooren 1999, 103). A 1982 study initiated by the Dutch Ministry of Housing, Spatial Planning, and the Environment further recommended that waste heaps be shaped as foreign objects in the landscape rather than blended into it. It encouraged a landfill form generated by the technology and process of dumping (Dooren 1999).

The most renowned landfill project in this category is Nancy Holt's *Sky Mound* (1985), which drew waves of public attention. Her design proposal for the 57-acre garbage heap in Kearney, New Jersey, commissioned by the Hackensack Meadowlands Development Commission (a quasi-governmental agency), integrated a working facility for the production of methane gas with a monumental earthwork and sanctuary for humans and wildlife. Although the project was not fully realized, it set a key precedent for landfill restoration.

A surge of landfill art

New laws governing solid waste and water and air quality, a scarcity of open space, and public pressure soon set more waste site projects in motion. Designers and artists were commissioned to work on public projects, and they joined teams of engineers and scientists. This leap marked a new era in the politics and art of the "garbage factor." The art critic Thomas McEvilley claims that landfill art emerged as a subgenre of land art. It focused attention on the basics of life—issues of body and consumption—and rejected the transcendentalism of previous artworks to soul and spirit (1999, 62). Landfill art reappropriated and sanctified the earth over the sky. This inversion is illustrated clearly in *Green Cathedral* (1987–96), by Marinus Boezem, in which the floor plan of the Reims Cathedral was demarcated with planted trees on a landfill in Almere, Holland (fig. 3.10), and later in Patricia Johanson's 1999 design proposal for the Nanjido landfill in Seoul, South Korea, in which the giant landfill is shaped to represent mythical animals traditionally believed to ward off evil. "The erection of artworks on top of the mound," McEvilley writes, "serves as a celebration of life's reinvigoration through death, and of the sanctity of the community's tradition as a system of feeling wrung from the experiences of the dead and the gone" (1999, 63).

In America the artist Mel Chin took an alchemist's approach, applying a scientific method to the recovery of a dump. Working with Dr. Rufus Chaney, a research scientist specializing in polluted soil and ecosystems, Chin tested whether certain plant species known for their absorption abilities could help detoxify Pig's Eye landfill in Saint Paul, Minnesota. *Revival Field* (1990–93) was intended to catalyze a method of cleaning toxins from contaminated sites while making a visible, sculptural statement (fig. 3.11). The works of American artists Ukeles, Denes, Holt, and Chin in damaged ecosystems were documented and exhibited in *Fragile Ecologies,* an exhibition curated by Barbara Matilsky in 1992 at the New York's Queens Museum. Compared to the Earthworks exhibition in Seattle, this exhibition presented the artists as healers and problem solvers and cast the land as a fragile ecosystem.

Ukeles's continued inquiry into dumps provided potent material for two exhibitions in the 1990s. In *Garbage Out Front: A New Era of Public Design* (1990), an exhibition and a symposium at the Urban Center of the Municipal Art Society of New York, she brought together a variety of projects by artists and an architect that represented and demonstrated the complex issues posed in modern society by waste management. Recognizing that practical information is essential to public consciousness and that the logic of engineering technology drives the landfill and gives it form, she lined the exhibition hall's stairway with a full-scale, thirty-six-foot high, cross-sectional mock-up installation of Fresh Kills landfill, complete with layers of clay, geosynthetic materials, garbage, soils, and gas and water pollution monitoring systems (Hartney 1990, 13). Intent on directing public design resources to waste sites and exposing the paradox and politics of landfills, Ukeles organized the show as a journey that began in *Regulatory Maze,* a labyrinth of lengthy and convoluted legal regulations, and ended with examples of practical approaches to waste management.

In *City Speculation,* a 1994 exhibition in New York's Queens Museum, Ukeles used a video and audio installation to represent the latest methods and complex technologies used to map, measure, plan, and monitor Fresh Kills as means of entry into that daunting place (Ukeles 1996). *Fresh Kills: Imaging the Landfill/Scaling the City* was intended to raise the question of whether high-tech imaging systems at both macro- and microscale bring us closer to a relationship with the land or distance us further from it (fig. 3.12). It conveyed the landfill in factual and practical, as well as humane and poetic, terms through images, documents, and maps—Geographic Information System maps, video simulations, macroscale satellite images, microscale images of microbes digesting waste inside the dump, written regulations, and lists of the functions of the people employed at the landfills. The roster of more than 500 professional employees resembled, according to Ukeles, "a university personnel list." It included crane and compactor operators, engineers, environmental monitors, policy analysts, chemists, geologists, landscape architects, planners, metalworkers, office administrators, sanitation workers, soil scientists, surveyors, restoration ecologists, and bird counters. The audio work was a composition of field recording of more than forty bird species that are attracted to the site and particularly to the marshes, including the more than fifty thousand seagulls that hovered over the active bank during dumping and rested at work's end on the hillside or the fences in the water. The birds attract avid Audubon Society birdwatchers for the annual ritual of spotting and counting the birds on the sacred and quiet day of Christmas.[8]

Ukeles calls Fresh Kills "the city's most comprehensive, democratic, *social sculpture."* She sees the place as a stage for dialogue "where you can ask a lot of questions and get answers," a kind of free public university of nature-human ecology in real time and space (Yung 1996, 28). For her, the waste system and facilities are defining and reflecting our relation to the material world. They could be our *"public cathedrals . . . our symbols of survival"* that can help us change the relations we have now (Ukeles 1995, 193).

The pitfalls and opportunities of postmodern dumps

Nearly half of the nation's more than seven thousand municipal land-fills were capped by the mid-1990s.[9] Hundreds more old dumps have been closed and new ones opened since then. "Landfill reclamation" entered the professional jargon and practice of design disciplines. Practical, ecological, symbolic, and formal considerations drove landfill reclamation during the 1980s and 1990s, but the focus was on closure and end use rather than on new ideas for landfills and waste disposal.

Greatly encumbered by external factors—soil settling and gas emission—that generally prohibit intensive development of landfill sites for several decades, the design of landfills limited itself to recreational uses, such as golf courses and parks. A slew of state regulations and engineering requirements to guarantee public safety, minimize erosion, and control pollution has created both a technical challenge and a major financial burden that consumes at least two-thirds of every project's funding and leaves little for design. Environmental engineers and ecologists devised rigorous scientific solutions and new bioengineering technologies to safely construct and close landfills. Environmental designers and artists, proposing to "open" them to the public both physically and intellectually, often encountered strong resistance. Only a few of those designing dumps pondered how technical regulations and the technology required can become an opportunity rather than a constraint.

Unfortunately, the dichotomy between technology and aesthetics and the traditional separation between engineering and design / art commonly keep the two tasks apart. Technology is perceived as a means to solve problems created by the dump; it allows but does not shape the new reality of the place. Also stifling creativity is the tendency to consider a site merely as an object, another commodity, rather than as a living, changing sculpture both during and following its operation as a dump. The real challenge of landfill reclamation is to incorporate construction and closure technology, and the ecological processes at play as part of the overall design language of the new place—as form-givers, at once utilitarian and aesthetically and intellectually satisfying.

Designers are also responsible for erecting obstacles to their own creative ideas. When imagining the fate of dumps, many designers (particularly landscape architects) and, to a lesser degree, artists often carry with them the burden of society's ecological despair and the agony over a lost paradise. Unable to overcome the oppositions between nature and culture or technology, their practice for the most part has been rooted in the idea of marginal places as nuisances that need to be fixed. The "dead," barren, ugly, polluted, and wasted site must be reborn as a "living" and useful place. Often embedded in these camouflage and restoration approaches are canonical landscape typologies, encoded with romantic ideals of wild or tamed nature, that mask the place's essence. Acting on their intentions to reconnect humans with nature, the designers shape and cap the dump with a green suit of grasses (with only methane wellheads attesting to its previous identity). Trees, once prohibited for fear that they would fracture the protective cap, have been shown to be harmless and now provide welcome aid in completing the pastoral scene.[10] At times, taking a utilitarian approach, the land is assigned some recreational use to recover its value and add a sense of communal worth. Likewise, many closed dumps became traditional parks for ball games, jogging, biking, and relaxing in scenic surroundings. Some were reformed as exclusive golf courses. Many contemporary dumps have embraced nature and become wildlife reserves (with or without an environmental center and educational programs). Others have become elegant sculpture parks. Consumed with technological guilt, these designers overlook new aesthetic possibilities presented by the latest technologies and waste disposal methods; they bury engineering solutions and operational procedures along with the garbage inside the dump.

The resurrection of dumps as pastoral parks and nature reserves, or alternately as recreational grounds, transforms the Other and unnatural into the familiar and natural. These approaches reessentialize binary oppositions from the side of nature and leisure, erasing the distance that is needed to look critically at our landscapes and habits. The critic Rossana Vaccarino has argued for the need to create places that go beyond simple recreation—places that establish a feedback process and question the very creation of waste. For her, the transformation of a wasteland into a park, natural or otherwise, is the ecological problem, not the solution. "Successful parks" atop a landfill cover up their exploitation, rendering it forgivable and pointless. She claims that in true sustainable design, waste must be recycled "as a raw material for new biomass" (1995, 91). Thus, the key questions remain: Does the new design seize the opportunity presented by the unique condition of a landfill? What message does the place finally convey? Should the landfill reclamation explicitly address the waste problem? In what way could the landfill benefit the nearby community?

While the public hails the transformation of landfills from places chaotic and dead to ordered and alive, those eccentric few who delight in the hidden intrigues of these places lament the erasure of their surreality, difference, and peculiar dynamics. The excitement of Robert Sullivan, the explorer of New Jersey's Meadowlands, vanishes when he encounters the only "fixed," grass-covered landfill in the area, which appears to him a bucolic, opaque, and benign landscape rolling down to the river (1998, 152). The risk has always been that in "cleaning" these landscapes and "solving" their real problems, we program into them our conventional tidy aesthetics and repress their allure, the qualities that made them attractive sites of exploration and adventure in the first place. In doing so, we end up flattening our landscape.

Insights into Dumpsites: Projects and Critiques

The prominent projects of the late 1980s and 1990s fall into four characteristic types. The first, represented by DeKorte Park in New Jersey, exemplifies the restoration approach, turning the surface into a natural oasis, the antithesis to the disgusting material below the surface. Danehy Park in Massachusetts typifies the second type, the utilitarian approach that transforms dumps into recreational arenas for play. The third type is represented by *Sky Mound* in New Jersey and an art addition to Danehy Park. These projects celebrate and ritualize the dump, considering it a kind of cultural monument and communal hub. In the fourth type, exemplified by Byxbee Park in California, the site and its context are opened for discovery and individual interpretations. Taking the celebrative approach, the last two groups endow dumps with identities that transcend, yet build on their former realities and expose complex and competing demands, regulations, interests, desires, fears, and dreams.

The natural hill: DeKorte Park

The prevailing agenda for landfill restoration is to return the site to "nature." Such an agenda guides the initial layout of the built earthform, which favors soft contour lines in order to blend well with the "natural" surrounding. Following the capping, the ground is seeded and planted with a green mantle, symbolizing a redeemed life.

One of the most celebrated landfill restorations is Richard W. DeKorte Park, 110 acres of mostly brackish marsh in Lyndhurst, New Jersey. Garbage islands, dredge spoils, dikes, and tidal impoundments were transformed from toxic sinks into a nature reserve and park. Located at the edge of the large Bergen Landfill (or Kingsland Landfill), the site was chosen to symbolize a stand against the continued destruction of the marshes by landfilling. It provides an ecological refuge for wildlife; environmental education for local residents and other visitors; a laboratory for ecologists, agronomists, and bioengineers; and a location for the headquarters of the Hackensack Meadowlands Development Commission (HMDC), an agency created in 1968 to oversee the reclamation and development of 32 square miles of degraded land in the Meadowlands. The commission's ecological agenda for transforming the damaged ecosystem into a biologically diverse habitat is manifest in the many reclamation and mitigation projects in the area, but critics claim that the projects are merely another type of development, this time of nature, intended primarily to improve the image of regional industries and of the state of New Jersey, in addition to meeting the needs of residents for recreational and open space (Pottieger and Purinton 1998, 225).

Since the late 1980s the agency's landscape architecture team, led by Katherine Weidel, has worked with natural resource specialists and engineers to reestablish a habitat that can be found under similar conditions elsewhere in the region. Weidel considers the approach adaptive and pertinent and explains, "Although some people would have liked to see the cedar forest ecosystem once dominating the area return, this was not a realistic goal" (conversation with author, Meadowlands, N.J., March 13, 1999). The contamination of the swamp by the area dumps and industries posed an unprecedented challenge: engineers tested biotechnology to close the landfill and stabilize its slope; scientists experimented with high-tolerance native plants; and landscape architects programmed use, interpreted ecological habitats, and made the park accessible via a series of paths and structures (some made of recycled materials).

The park opened in 1990. It now attracts 250 species of birds and offers a variety of public programs and activities that celebrate the redemptive act of nature's resurrection. The park's designers emphasize that the rules guiding the design are not those of a traditional park but rather the ecological principles of nature, as they aimed to replicate the natural succession of wetland and upland habitats. According to a park brochure, "Plants are arranged as Mother Nature would have planted them, in hedgerows, thickets, and a mosaic of meadow grasses and wildflowers" (HMDC 1992). But the park, which combines four distinct areas, makes little effort to reconcile intent, style, and vocabulary. The Kingsland Overlook is a controlled and tidy arrangement of a "natural" succession of landscape; Garbage Island is replete with suburban park types of plants and structures (fig. 3.13); the marsh is a dense wetland for wildlife observation; and Transco trail is a linear path laid on top of a gas pipeline service dike.

The park is structured along a series of trails that accentuate the formal quality of each section. The marsh boardwalk and Transco trail seem to follow the logic of the site and to highlight the experience. The two paths create a loop anchored by the park's centerpiece, the HMDC environmental center—a corporate-looking building perched on the water. The marsh trail follows the curves of the old Kingsland Creek, juxtaposing the intimacy of the marsh and the industrial transmission lines and freeways of the surrounds. Transco trail, the sole technological vocabulary at the park, traces a straight gas pipeline service dike and is lined up with rhythmic yellow pipes that mark the gas line below. Conversely, the curvilinear alignment of the trail systems on Kingsland Overlook and the Kingsland Nature Reserve—the two reclaimed garbage islands—seems more arbitrary and picturesque in nature. The park yields to its overpowering surrounds without entering a dialogue with them. The overall design scheme is at minuscule scale, which does not respond to its context—to the surrounding landfills, industries, and infrastructure facilities. Moreover, the innovative technology used to construct the park does not add visibly to its texture, despite the docks, the constructed access to the water, and the PVC viewing tubes that frame regional monuments.

The design details, patterns, structures, and interpretive elements are confined to an internal and singular agenda, nature. References, signs, exhibits, and names are replete with words like "nature," "native," "natural," and "life." The rich human history that shaped the area—the route used to transport ore in the colonial era, the railroad dikes, and the gas industry—is omitted in favor of natural history. The Transco trail, for example, could have been a place to highlight the site's cultural ecology. This section of an important 1,800-mile gas pipeline—leading from the Gulf Coast, off the Texas and Louisiana shores, past the site and to storage tanks in Carlstadt, New Jersey—serves 80 million people in the metropolitan area but is denied clear interpretive acknowledgment.

Likewise, the Hackensack Meadowlands Environmental Center is dedicated to nature's ecology. The building is oriented toward the restored brackish marsh, turning its back to the large Bergen Landfill. A diorama of the early cedar forest ecosystem welcomes the visitor. Picture panels depicting the transformation of the garbage islands into a renewed ecosystem are displayed at the entry to section that deals with the garbage "problem." It invites visitors to see and interact with various garbage displays and visual simulations and uses colorful cartoon characters to tell the garbage crisis story and its possible solution, but it does not suggest tours to the many open and closed dumps and garbage facilities in the area.

The ecological agenda of DeKorte Park is important but weakened by its one-sided presentation. A more effective approach would be to display the human dimension and technology as part of nature, with the park's context integrated into and reflected in the site.

The recreational arena: Danehy Park

A widely accepted alternative to reliance on the natural scheme and extensive nature-related activity is an urban park scheme that turns the waste site into an intensive public recreation amenity. In more urban environment, different community groups, each with its own agenda and recreational needs, demand a share in future gains.

One well-known and popular landfill park is in Cambridge, Massachusetts. Mayor Thomas W. Danehy Park is built on a 50-acre former clay pit and brickworks that later became a municipal dump. Monitoring and testing of the site followed the end of active dumping in 1971. In the late 1970s the Massachusetts Bay Transportation Authority (MBTA) dumped the excavation material from Boston's new red line subway extension there, compacting and covering the municipal waste below. It created a 40-foot mound on top of the 70 feet of buried garbage. The dump's location at the northwest lowlands of Cambridge (not far from Fresh Pond, the city's reservoir of drinking water), the landfill gases, and flooding of neighboring areas became major public health concerns and finally forced the city to embark on the reclamation project. Opening in 1990 as an all-encompassing urban park, it added to the densely populated community 20 percent more open space, loaded with ball fields. Describing opening day, the reporter for the *Cambridge Chronicle* exclaimed: "The sense of wonder was unanimous. The disbelief was shared by all. There was a feeling that miracles can happen, that anything is possible" (Gaulkin 1990).

Lacking any earlier precedents to follow, the project was an engineering, public works, and design challenge. It began in the early 1980s amid skepticism and suspicion. "People could not or did not believe the park could happen," says Deputy City Manager Richard Rossi, who worked at the dump as a teenager and still remembers its "evil and dangerous" ambience (interview with author, Cambridge, Mass., March 18, 1999). Defusing public fears and accommodating the desires for recreational facilities were the first tasks facing Rossi and John Kissida, a landscape architect and co-principal of the engineering firm Camp, Dresser & McKee Inc. Kissida, who loves designing golf courses, admits that he did not like the hard edges of the mound and wanted more of the natural shapes associated with parks (interview with author, Boston, March 18, 1999). His key concern, however, was technological. Kissida devised an accomplished infrastructure for the park—systems to control erosion, guarantee proper drainage, and monitor gas and water pollution (Kissida and Beaton 1991). Yet, the technology that made the park possible did not inform the design language. While one conspicuous element is a wetland habitat for stormwater retention, wildlife, and nature study, the peripheral vent trench and pollution monitoring wells that retain the stormwater are hidden (though they are described in the park's informational brochure). To accommodate the many desired ball fields and recreational programs, the designer lowered and shaved the top of the mound into a plateau, thereby taming the views to and from this human monument.

The park is saddled with a mosaic of ball fields. A football field and a small amphitheater with a painted lily pond at its asphalt center stage were recently added to the already overloaded scheme.

Ritual ground and monument: Danehy Park revisited and *Sky Mound*

Despite the crowding, a special project was given space within Danehy Park. Mierle Laderman Ukeles chose the site for her Cambridge Arts Council Art Insite public art commission in 1990. Her intention was to evoke and heal the memory of the dump and open up the site to celebrations and rituals. Ukeles has always considered mounds, including garbage mounds, as sacred sites and hopes to direct attention and bring people to the top of the hill (interview with author, New York, March 16, 2000). *Turnaround Surround for Danehy Park* recalls the lost identity of the dump, centering on its transformative powers and relevance for the community. Ukeles chose the top of the mound, the highest place from which to see into the distance, to site her artwork. The physical intervention includes four parts, two to be built in 1993–94 and the other two in 2002–3. The first element, the "glassphalt" path, creates access to the top of the mound and then descends back down (fig. 3.14). It embodies the idea of a turning point. The path is made of crushed colored glass and mirrors that were donated by schoolchildren and local companies, as well as materials collected by the city recycling program; contrary to most functions in Danehy Park, it is accessible and inviting to *all* people. The glassphalt glitters in the sun and becomes more beautiful as it wears down and ages.[11] The second element, the "Smellers and Wavers," comprises herbs, roses, and grasses planted along the path. They accentuate the wind, indicate the return of a healthy ecosystem, and emit fragrance in a place that used to smell bad.

The last two elements will include two circular discs of 20–24 feet in diameter, also at the hilltop. The first, the "Throne Room for the King and Queen of the Hill" will be made of molten, recycled rubber and tires embedded in concrete. The second disc named "Dance Floor" is to be made of layers of black recycled rubber bits bound by a transparent medium, with a galaxy image of colored rubber and surrounded by concrete seating blocks. The discs are to become spaces well suited to group gathering, dancing, and play. "Community Implants," the last element, will include several containers embedded in special niches within the Throne Room structures. These containers will enclose precious objects, "offerings" brought by community residents of different ethnic backgrounds and will be etched with quotations. They act as counterpoints to the garbage below.

But the physical interventions are only one part of Ukeles's artwork, which also relies on performances, events, meetings, and lectures to educate people about solid waste and nurture their participation in and ownership of the park. Ukeles explained her ideas about the park to the community, schoolchildren, and the city, and she organized glass collection events. Both Kissida, the park designer, and Rossi, the deputy city manager, value Ukeles's vision. Kissida points out that it is easy to forget the park's history: "People are upset when they cannot play ball on the fields after a rainy day. They forget it was a dump" (interview with author, Boston, March 18, 1999). Ukeles is interested in making Danehy Park meaningful to its community by reminding them of the dump that was there, celebrating that reality by inventing new community rituals, and making them conscious of the ways they are related to the material world.

An earlier art project on a Meadowlands dump is also celebratory and ritualistic, even though it was not fully realized. Unlike Ukeles, who had to work within an existing, inner-city park, Nancy Holt took the liberty to mold her outskirts dump into a monumental earthwork and program it as she desired.

Sky Mound, a 57-acre dump in Kearny, New Jersey, not far from DeKorte Park, was the first public art project on a landfill commissioned by the HMDC in 1984. The 100-foot-high earth monument, tightly bounded by transportation routes and transmission lines, is seen annually by millions of people who pass through the area by car, train, or plane; it serves as a gateway landmark to New Jersey. Holt's concept derived from her ongoing preoccupation with marking and celebrating time and sky, as well as from the historical precedent of American Indian mounds. "Both kinds of human-made mounds were built to meet vital social necessities, but here the similarity ends," Holt observes (1995, 61). She sought to bestow on the modern garbage mound some of the spiritual, social, ritual, and transformative powers of its ancient counterparts.

Phase one of *Sky Mound* (currently named I-A landfill) is complete. Holt worked closely with the engineers to merge the landfill closure measures and the design language. The garbage heap is sealed and capped properly; the underground methane gas recovery pipe system and the methane wellheads, arched in accordance with the design, are in place; and on top of the mound a small pond that offers a habitat for wildlife, retains rainwater, and reflects the sky is finished. Wild grasses and flowers have been seeded to provide erosion control and wildlife habitat, and the groups of metal pipes protruding from the ground are positioned to

visually align with celestial bodies (the sun and moon) on astronomi-
cally significant dates. But the site is closed to the public, and work on
the project has been halted.[12]

The unrealized proposal shows two areas on top of the mound, lunar
and solar. They mark two formal centers from which radiating sight lines
of widening gravel paths project. Framed by pipes and earth mounds,
the paths run down the slopes and serve as a visual-formal armature for
the mound (fig. 3.15). On equinoxes and solstices, the sun would be seen
rising and setting directly over the center of the solar area. The rising and
setting of the stars Sirius and Virgo would be registered by tunnels and
stairways in two of the mounds, and the extreme positions of the moon
would be aligned with the arching pipes. Although the garbage is concealed with a durable, synthetic rubber liner, the measuring devices and
the burning flames of methane gas lighting the skyline are intended to
make the underground garbage "visible" (Lovelace 1987; Matilsky 1992).
Other vertical and horizontal devices are designed to amplify the
processes and the particularities of the landfill, as well as its connections
to other places. Spinning wind vents, similar to those used to vent landfill gases, are to be placed on the lower slopes of the plateau, and a steel
measuring pole, whose top marks the original height of the landfill summit, is meant to reveal the gradual settling of the mound as organic matter decomposes. In addition to the vertical connections—sky, people,
and garbage—horizontal connections would recall other distant landfills in the country.

This program emanates from the site's monumentality, location, and
dramatic encounter between earth and sky. It is designed as a place to as-
cend, a place from which to see to the distance and to be seen by as many
people as possible. It is intended as a destination for rituals marking time,
seasons, and space. "These heaps of rubbish," Holt declares, "will be seen
as the artifacts of our generation, our legacy to the future. So there is no
escaping our responsibility for making these mounds of decaying matter safe by using the latest closure technology, and eventually reinter-
preting and reclaiming them, giving them new social and aesthetic
meanings and functions" (1995, 63). Holt's 1985 design was an original
response to the dialectics between technology and nature. The awesome
feeling of the site is accentuated by the magnitude of scale and the bold,
formal strokes so appropriate to this kind of place.

Exploratory terrain: Byxbee Park

Like *Sky Mound,* Byxbee Park in Palo Alto, California, registers the wind, sun, and water at the site, but with more subtle gestures and formal impositions. It remains a place for individual discovery rather than group gathering and rituals. It avoids playing fields and "natural-looking" design, yet it is both natural and recreational. Those aspects are developed from the dialectics inherent in the site and its context, in ways deliberately artifactual and contemplative.

The artists Peter Richards and Michael Oppenheimer collaborated with the landscape architecture firm George Hargreaves and Associates on the master plan for the 45-acre Byxbee Park (a portion of a 150-acre landfill). The landfill is located at the edge of the San Francisco Bay and fragments of a marsh, surrounded by many infrastructure and industrial sites—a municipal airport, a cluster of transmitting antennae, a recycling plant, a not far-off sewage plant, a golf course, and a freeway. In 1987 the Visual Arts Council asked the design team to work with the Public Works Department and the citizens of Palo Alto to rethink an earlier master plan for the landfill. Following an intensive period of site visits and discussions, the team decided that "it did not want to hide the site's origin as a dump but to have the context to inform the design of the park" (Oppenheimer, phone interview with author, May 6, 1999). They aimed to create a design focused on the interplay of human-made and natural processes.

A small budget (one-third of the total closure budget) and uncompromising closure standards forced the team to accept a kidney-shaped hill that had formed based on an earlier plan, which was possibly intended as a golf course. The team proposed to create a place for public art and for extensive use as a park. "We wanted visitors to have a journey of discovery, to try making sense of the funny, weird sculptural elements in the park, as well as make physical and cultural observations about the area," says Richards (phone interview with author, February 26, 1999). Team members used subtle earth sculptures and artifacts to evoke the memory of the buried content. The sealed waste takes the shape of conspicuous artificial hillocks and swales. The hillocks, a main feature of the site, recall Alone Indian shell mounds found along San Francisco Bay, create spaces to hide from the fierce winds, and provide dramatic vistas. The combination of prospect and refuge also affects patterns of moisture and redirect wind. At the top of the mound a methane gas collection unit dominated by a dramatic flare is visible at night; its shadow can be seen on a keyhole-shaped white gravel bed during the day. The shape of this gravel bed playfully suggests that one is peering into the secrets of the landfill, whose essential core is a decomposing bed of trash (Rainey 1994, 176). Like other dispersed visual references and sculptural gestures throughout the park, the meaning of the gravel bed is implicit, layered, and not readily accessible.

Fragments of the industrial culture surrounding the Byxbee landfill are conspicuously integrated into the grassy park as sculptures, as signifiers of the surrounding infrastructure and amplifiers of wind, time, and change. For example, chevrons made of portable concrete highway dividers are positioned curiously, running down the slope in alignment with the municipal airport runway; they signify "do not land here" in aeronautical symbols to pilots descending to the nearby airport. These artifacts merge ecological, aesthetic, and symbolic dimensions as they create bold patterns against the finely textured grass, retard erosion, nurture pockets of wetness for the moisture-tolerant plants, and reference elements at the distance. Relying on a similar logic, a field of seventy-two electric poles, echoing those in the distance, is set in a grid with a tilted top plane (fig. 3.16). The poles accentuate the topography of the slope and together act as a giant sundial. The ground's subsidence registers in the poles, which lean together drunkenly. The critic Rossana Vaccarino sums up the project: "There is no one privileged signified and the play of signification has potentially no limits. The park's score can be played in many different ways, and the visitor is a sort of co-author of the score who improvises the music rather than giving it a true expression" (1995, 89).

Functional elements composed of observation platforms built by the shoreline serve the solitary jogger and bird-watcher, and walkways covered with crushed oysters shells allude to the shell middens left by inhabitants hundreds of years ago. The white paths provide a contrast with the green grasses that cover the landforms. To cut down on maintenance, hardy exotic rather than native grasses were planted. Byxbee Park has retained a sense of open-endedness and casualness. It reveals the discursiveness inherent at and around the site. Regrettably, however, the design makes no intentional gesture to integrate an adjacent operating landfill within view of Byxbee Park.

Another dump project designed by Hargreaves Associates, not yet implemented, molds the 470-acre Mont-Saint-Michel in Montreal, once a limestone quarry and later a city dump, into a center for environmental education, recreation, and performance. A series of formal configurations and spaces draws attention to the place's own morphology and spatial qualities—tall, ragged quarry cliffs and ledges and soft sloped volumes of waste fill. The design draws on and juxtaposes itself with the history of excavation and filling, as it continues to sculpt earthforms that now fall within predetermined technological specifications and constraints—slopes for drainage and berms for biogas collection (Hargreaves Associates 1997).

The landfill park is a deliberate composition of discrete spatial-geomorphologic parts—the playing fields, the labyrinth, the amphitheater, the wind field, the belvedere, and the quarry field. Each creates a distinct space and experience. For example, the labyrinth features raised, angular berms planted with meadow grasses and built around the site's network of gas collection pipes and wellheads. The wind field features curving swales that direct people and drainage to a central pool in the lower quarry area, in which conical earthforms, made of rubble excavated from the quarry floor, stand erect like old quarry tails (O'Connell 1999, 48). The formal and stylish design could have been complimented by a less programmed and more open-ended area, as well as bolder references to the chaotic nature and informal history of the dump.

The above four types—natural hill, recreational arena (ball field), ritual grounds and monuments, and exploratory terrain—represent a range of postmodern approaches to landfill reclamation. They draw on history, community, symbolism, and experience. Yet, new landfill "reclamation" projects are beginning to take shape across the Western world. Deprived of traditional visual, formal, and symbolic dimensions, they address the waste management issue head-on, formulating viable, supermodern aesthetics for dump design.

Dumps of the Future: Exploratory Scenarios

The Dutch landscape architect and journal editor Noel van Dooren declares, "The idea of a garbage dump as being something that eventually reaches saturation and thereafter must be given a new use and meaning is undergoing radical change" (1999, 104). Partly because they need costly and almost perpetual monitoring and maintenance, landfills will be avoided in the future. In Europe, the output of the waste has already been reduced; a good deal of it is recycled, another part incinerated, and only a fraction buried.[13] Demolition waste, in particular, is sorted into heaps of wood, iron, aluminum, glass, concrete, and so forth and then shredded or processed in new "dumps" that serve as transit points (104).

Dooren identifies three directions for future dumps that epitomize the spirit of supermodernism. Two prospective types were winning proposals in a 1994 competition sponsored by the Dutch Association of Waste Processors. In their entry, *Megastratum,* the engineering firm DHV considered the dump as a temporary storage space where each kind of waste material is piled in a specified mega-pyramid tank until appropriate processing and reuse methods are found. The second proposal, *The Recyclable Dump* by Grondmechanica Delft, conceives of the dump as a processing factory where entering materials are processed via sophisticated bio- and geochemical conversion and readied for transport to future use.

A third approach concentrates on the future of closed dumps and their potential as resources for excavation of raw material. Landfill mining has been experimented with since the early 1980s. Excavators, screens, and conveyors recover buried materials suitable for a variety of purposes, such as composting, use as construction subsoil fill, combustion, and recycling. Among the reasons for these experiments are saving landfill space, remediating dangerous areas, and upgrading landfills to meet environmental regulations. As of 2003, six landfills had been mined in the United States; several similar operations are known to be under way in East Asia and Europe.[14] It is generally necessary to wait fifteen years for material to stabilize and decompose, although even then quality fluctuates depending on what has been buried and the methods of landfilling that were used. The economic future of landfill mining is not clear. Nonetheless, in these future scenarios—as a place for storage, an arena for processing and recycling, and a repository of resources— dumps as permanent land features and burial sites might not exist.

The recyclable dump

When thinking about landfills involves rethinking the creation of waste, not just its disposal, then the landfill becomes a place for researching and investigating new practices and ideas, a place that re-creates the dump not for leisure activities but for new social, biological, and economic functions. The site itself is not recycled for a new use; rather, it enters into a continual feedback process in which waste is recycled on-site as raw material for new biomass and people actively envision the production of objects and new lives for them. These working centers are designed as elegant and flexible structures, subject to stringent laws that protect the environment and nearby residences, and offering employment and other programs to the near community.

Some closed landfills "naturally" turn into recycling and processing sites. Private corporations build their enterprise, with its new mission, on top of the old landfill. The waste stream continues to arrive via the same route, but instead of being buried, it is resurrected. Regional composting centers already operate at closed landfills around the United States (e.g., Des Moines, Iowa, and near Philadelphia). For other landfill sites, cities or counties are devising plans for complexes that integrate research, learning, and enterprise, all focused on reusing and reducing waste material.

The recycling complex Sun Valley Recycling Park, by Waste Management Inc., is awaiting realization at the 205-acre Bradley landfill near Los Angeles, California. Various materials are already being bailed and processed at the site, and methane is being captured to fuel an electric power plant. The elaborate future plan features a dozen recycling facilities, whose tasks will include composting, soil remediation, demolition, processing green waste, processing tires, milling paper, and recovering building material. A rail haul facility will help move resources in and out of the site, and the landfill gas will provide fuel for vehicles used at the site. Recreational structures, vegetation, and trails will be integrated into the park complex to ensure its viability and its use by the community (Riggle 1995).

An ambitious master plan, named the Center for Environmental Learning and Enterprise (or "the campus"), is currently being considered for the Phoenix, Arizona, 27th Avenue Landfill, which is adjacent to the 27th Avenue Waste Transfer and Recycling Facility (see chapter 4) and the 23rd Avenue Wastewater Treatment Plant (see chapter 5). The blueprint, which builds on and feeds off Phoenix's existing waste facilities, aims, in the words of its designers, to create "a new type of urban open space—a place of environmental learning and enterprise—and, in doing so, transform a degraded parcel of land into a place of renewal and productivity" (Glatt et al. 2000, 7).

The landfill, almost a mile long and 70 feet high, makes up the northern third of the 407-acre campus. It abuts a large middle zone, an area where the sludge ponds of the sewage plant were formerly located and the recycling plant is now placed. The southern third abuts the Salt River channel and its degraded riparian strip. A collaborative team of two artists, Lennea Glatt and Michael Singer, and two landscape architects, Fritz Steiner and Laurel McSherry from the University of Arizona, was charged by the Phoenix Arts Commission and the city Public Works Department to create a vision for the area in the spirit of the existing waste facility's commitment to environmental education and its community agenda. The program proposed for the area combines mulching and recycling operations, environmental research, and an eco-industrial park with workshops, studios, classrooms, galleries, demonstration areas, and gardens (McSherry and Steiner 2000). The footprint and architectural details of the existing waste facility directs the campus's diagonal layout and the facilities' future formal vocabulary (fig. 3.17).

The landfill will serve as a main gateway to the campus, offering standpoints from which to survey the area and providing panoramic views of the city. Pedestrian and cycling paths will thread the garbage hill, connecting site overlooks, interpretive centers, and a landfill research area where landfill mining will take place. The design of the landfill seeks to draw on and connect to the layers buried at the site. The planned lower research and enterprise zone is a complex of structures, recycling yards, and aquaculture and nursery facilities all designed to function in a closed loop. The eco-industrial functions depend on one another's byproducts and recycled products, make intelligent use of existing land and climatic resources, and operate to improve the local landscape and advance environmental and waste management research and industries. To accomplish the goals of community learning and environmental education, a vocational charter school will be established, as well as several interpretation and demonstration areas focusing on water and solid waste resources in the desert.

The campus will also serve as a gateway to the Rio Salado restored greenbelt system, which will receive water recharge from groundwater and from the effluent canal that bisects the site. Open spaces and gardens interspersed throughout the area will house desert plants and provide a welcoming space for people. "The project," state McSherry and Steiner, "is motivated by our desire to make vivid what could be thought of as the *ecology of place*" (1998, 3). In other words, the team has tailored a design in tune with the site's own identity and its ecological and economic capacity. Rather than being an end point, the landfill becomes raw material for use and for generating future waste practices. It partakes in an ongoing exploration of the relationship between waste and society (McSherry and Steiner 2000). State and federal funds for the ecological park and restoration of the river corridor have been secured. However, securing the integrity of the plan's recycling agenda and luring the businesses that could make it function have proved a more difficult task, especially as the city caves in to pressure for adapting the land for a variety of other uses.

Two dumps, two art exhibitions

Unlike built projects, which must withstand the complex reality of regulations, economy, bureaucracy, individual egos, and community desires, hypothetical projects are free to explore, question assumptions, break taboos, reveal limitations, and suggest unexpected directions. The dump has long been a stimulus for freethinkers. Intellectual and artistic inquiries have always transcended the place and content of the dump to tamper with deep-seated cultural constructs and praxis. Through creative visual representations, artists and designers stimulate dialogue, propagate ideas, and alter common perceptions. Two such theoretical investigations of landfills are seen in the conceptual proposals of art exhibitions: one for the Hiriya landfill in Tel Aviv and the other for the Bavel Landfill near Breda, the Netherlands.[15]

The exhibition, *Hiriya in the Museum: Artists' and Architects' Proposals for the Rehabilitation of the Site* (November 1999–February 2000) took place at the Tel Aviv Museum of Art and consisted of nineteen proposals presented through drawings, models, videos, and installations (Weyl 1999) (fig. 3.18).[16] *Tales of the Tip: Art on Garbage* (May–October 1999), occurred at the Breda landfill site; it included five temporary art interventions and seven conceptual designs presented on billboards located along a route surrounding the landfill, just as billboards announce prospective construction sites (Driessen and van Mierlo 1999) (fig. 3.19). The two exhibitions stirred debate about the design potential of dump sites and what visual artists and architects could contribute to the spatial treatment of former dumps. As the proposals contemplated the "problems" and the potential of the sites, as well as the peculiar marriage between art and garbage, they also touched on related issues of waste, land, and culture—habitation, consumption, leisure, landscape aesthetics, technology, success, and self. Ideas ranged from the established and popular to the critical and eccentric, from the serious and constructive to the cynical and subversive. Several proposals clearly questioned the very assumptions underlying the exhibitions. Considered together, the two art endeavors reveal recurring responses and trends.

One approach evident in both exhibitions displays tendencies toward naturalist romanticism, restorationist correctness, and utopian dreams that are often inseparable from the impulse to turn a landscape into a commodity—a theme park where nature, animals, and garbage are packaged as a weekend's entertainment, complete with opportunities to spend money. (This approach was more apparent in the Hiriya proposals.) For example, the proposed ecological theme parks of the American artist Meg Webster and of the Israeli artist Erez Rota Sishoka both cram the Hiriya mound with extravagant, all-you-can-take-in experiences of nature, garden, and park. Webster's "Proposal for Hiriya Landfill" is bursting with a hodgepodge of landscape prototypes and recreational grounds—botanical and water gardens, forests and waterfalls, canals, ponds, islands, mounds, mazes, marshes, canyons, golf courses, cable lifts, restaurants, and, of course, composting areas, filtration systems, and educational facilities. "The Bird Park," by Sishoka, is also an environmental-ecological purification site, where water purified from the creeks below the mountain will emerge on top as fountains, waterfalls, and lakes, thereby adorning the myriad recreation uses and gardens proposed for the site.

More elegant and reserved, yet similarly shallow and trendy, are the Hiriya proposals of the Israeli landscape architect Shlomo Aronson and the Israeli team of planners and architects Ulrik Plesner, David Guggenheim, and Mordechai Kaplan. Aronson proposes to create a park titled "Birds on a Pincushion," an aviary in the middle of the mountain, as a reminder of the many birds that came to feed on the garbage. Alternatively, he suggests a wild safari park called "A Green Island in the Dan Zone" as an extension of the adjacent Ramat-Gan Safari. In either plan, the mountain again is transformed into a commodity, an attraction framed by borrowed and palatable imagery that covers up and avoids the more cogent site-specific issues of garbage and consumption. The impulse to turn the place into "a beautiful giant nature sculpture," as proposed by Plesner, Guggenheim, and Kaplan, simply covers the mountain with year-round bougainvillea plants bursting with red blooms, carves an amphitheater at its northwest corner, and surrounds it with grand lakes (Weyl 1999, 105). The resultant "natural-looking" landmark tames the extraordinary and creates amnesia about the buried repugnant matter.

Displaying utopian, modern, futurist tendencies are the two tantalizing proposals of the Acconci Studio, led by the renowned multimedia conceptual artist Vito Acconci—"Garbage City" for Hiriya and "A City That Rides the Garbage Dump" for Bavel. In the tradition of architectural visionaries like Paolo Soleri, Acconci transforms the wastelands into sanitary, self-sustained, and energy-efficient cities. In Acconci's proposal for Hiriya, "A city from garbage, with garbage, and for people," people live, work, learn, and relax on the garbage mountain (Acconci Studio 1999, 51). The slopes of the dump are stabilized by concrete mesh, an open-grid retaining wall that doubles as a building with plug-in private houses, schools, and offices. There is a stadium, a convention center, a public park, and fields for cows to graze and for crops to grow (fig. 3.20). The garbage below powers an energy plant, and marshes clean the contaminated water around the mountain. This fantastically intricate proposal, which is displayed in model form and accompanied by detailed engineering drawings and Acconci's mesmerizing descriptions of life in the city, is summed up in the very last sentence of his text: "You live off your city; you never have to leave home" (55).

For Breda, Acconci avoids fixed walls and structures, resolving to try new technological solutions. The dump is transformed into a city in the form of flying carpets made of saucer-shaped forms, which contain all the urban necessities. In "A City That Rides the Garbage Dump," the entire dump is paved with stones under steel mesh, and captured methane gas powers the city lights and the saucers floating above the landfill. Each rug is a web of bowls dedicated to buildings, water, or green environments (Driessen and van Mierlo 1999, 66). Acconci's highly theoretical proposals blur the boundaries between absurd and sensible, nightmare and dream, all the while effectively forcing attention.

Those who bypass the fantasies, the consumerist thrust to commodify the dump, and the moral preservationist mission to resurrect nature or ecology as a cultural panacea work to expose the limitations of these tendencies. This includes several proposals that provide a critical postmodern counterposition but no solution. The insightful Hiriya proposal of the American artists Mark Dion and Nils Norman does not offer a solution but rather lucidly expresses and critiques two common design tendencies (embodied in the proposals above). In "The Utopian and the Dystopic" two mannequins and two commercial billboards that advertise two contrasting theme parks, two impossible fictions: "The Hiriya Green Park" and "The Hiriya Hotzone: A Biohazard Extreme Theme Park

Experience." The fantastic, utopian park is represented by a "green guer-
rilla"—a white male dressed in camouflage military pants and vest and
a white T-shirt inscribed with a Hiriya Green Park logo. The hellish, cat-
aclysmic, dystopic park is represented by a rat wearing a protective uniform and gas mask and holding a Geiger counter. The "good solution"
miraculously transforms the site into an ecologically sustainable and educational Garden of Eden where visitors encounter environmentally
friendly facilities. The "bad solution" offers a park of extreme, hazardous
experiences, where, after signing liability waivers and donning protective outfits, visitors can take rides or stroll in mutated and toxic environments. Both parks are similarly self-deceiving, perpetuating the prevailing uncritical and absurd social desires.

Several other proposals refrain from presenting any design solution.
They prefer to cynically critique society or deliberately defy the assumption that artists should help direct the future of refuse sites. The Israeli team of the artists Gal Weinstein, Shai Weinstein, and Gil Vaadia
presents an intriguing and witty work titled "Hiriya: A Catalogue of Products." Claiming that art should not be used for healing, that artists
should not be considered saviors or Hiriya a wound, and refusing to
transform Hiriya once again into a commodity, they deal with the image of Hiriya instead of the site itself. They propose a consumer catalogue
of ordinary domestic products. Each is adorned with a Hiriya logo—a
fresh item of garbage, a garbage truck, or Hiriya's own silhouette—that
slyly earmarks it, too, for future inclusion in some dump site. Their installation is a commercial showroom: a dining room and living room setting complete with furniture and domestic items from the catalogue. The
artists have also drafted the store at the Tel Aviv Museum of Art as an assistant in their project: it sells shirts and baseball hats from the catalogue.

The French artist Hervé Paraponaris also uses the site as a platform for
critical thinking or, more precisely, illuminating the problem. "Dance on
the Volcano—Redeeming Como," a proposal for Bavel, is a platform that
invites people to rejoice with picnics, concerts, and fireworks while producing their garbage. The piles of garbage are then collected in Como
brand garbage bags and dumped into the landfill. Paraponaris programs
the site and conditions that represent society's desperate dance at the
brink of the volcano.

The Swiss team of Marcel Biefer and Beat Zgraggen (Bavel) took advantage of the Bavel site to proclaim as waste all of the art they had ever
created and dump it. They ask, Why should we even have art at such a

place? "Not only do they denounce artistic practices in an incomparably radical manner," observes the critic Chris Driessen, "but they also transcend the assignment of conceiving an intervention for the refuse site Bavel in which visual art is to serve as a starting point" (Driessen and Van Mierlo 1999, 27). The Israeli artist Igael Tumarkin's proposal "Hiriya as the Absurd" is a bitter criticism of the futility of reclaiming "Mount Sisyphus," as he calls it. Tumarkin sarcastically proposes to cover the mountain with a stabilizing colored net and to gouge deep, narrow canyons walled with painted concrete. These will resemble runways like those at the nearby airport. Subverting the curator's slogan, "Hiriya shall not rise again," he proposes that "Hiriya shall not fall again."

Another postmodern approach, more constructive and positive, seeks transparency and works to uncover and activate the latent potential of the dump to become a place of knowing and healing. Mierle Laderman Ukeles uses the Hiriya site to examine the state of Israel's environment. She devises a center for information and a stage for conversations about new environmental initiatives. Her proposal, "Evapotranspiration: This Land Lives and Breathes," shapes two temporal scenarios, day and night. During the day, four geysers spew mist from the corners of the mountain. Specific colors represent the state of health of the site and of the country's air, water, earth, animals, and plants. At the base of the geysers, stations equipped with electronic screens provide environmental images and data transmitted from the four regions of Israel. A table whose length (200 feet) equals the mountain's height is installed on top of the plateau and provides a place for six hundred people to gather and discuss their ideas for healing the country. At night, lights powered by methane and solar photovoltaic batteries bathe the entire mountain in colors reflecting the mound's own health.

The Austrian artist Lois Weinberger treats the dump as a cultural repository and a cultural mirror. "Present Time Space—Hiriya Dump" programs the site quite minimally and sensitively. Instead of a park that beautifies and glosses over everyday annoyances, Weinberger encourages the growth of seeds accidentally dropped by birds and wind and invites people to create their own paths and playgrounds. He proposes an elongated, rectangular-shaped, transparent metal structure for the top of the mountain. It will serve as a museum displaying exhibits retrieved from the buried mess, free of associations and full of contradictions. Weinberger's own display consisted of such retrieved items as bar mitzvah pictures that he developed from film he had found, newspaper articles, and a tourism video of Switzerland. His work captures and exposes a genuine and surreal portrait of culture and demonstrates the peculiarity and particularity of the site.

My own proposal, "Re-claiming Metaphors out of the Dump, or Four Gestures for Hiriya," engages four dimensions of the site: ecological (toxic liquid and gases), cultural (meanings and perceptions), engineering (structural stability and anatomy of the site), and visual (monumental presence). To access Hiriya's inherent multiplicity, I design gestures based on four metaphors—mausoleum of entropy, museum of rejects, archive of subconscious depositories, and monument of unpleasant necessities. Each metaphor is conceptually and literally dug out of the dump and redeemed, thereby conspicuously reconnecting Hiriya with its cultural and natural systems. The first gesture releases the captive processes of decomposition from inside the entombed mausoleum of dead commodities. Seven carved fountains at the foot of the mound spew garbage juice (leachate) into a series of marshes, "green kidneys" woven with a walking path to enable visitors to view the cleaning operation (fig. 3.21). As a museum of rejected material, Hiriya is connected through a second gesture to other museums of art and culture, to which it is inversely related. People whose garbage built the mountain are invited to extract some, encase it, and place it in front of their own cities' museums. As an archive, Hiriya is reconnected to its deep layers of deposited histories and geological grounding. New archaeological and engineering drillings to study and stabilize the mountain join existing hydrological test drillings in the area. The emptied cylinders are filled with reinforced concrete to stabilize the mountain; on top, derricks of imaginative constructions serve as props for special lighting and events. The built monument, the fourth gesture, draws connections and enhances high-level conversations between Hiriya and other contemporary monuments in its urban surroundings. A skyline promenade equipped with giant telescopes directs walkers and joggers to view other urban monuments and garbage mounds in the surrounding area. Each of the designated monuments provides information about Hiriya and redirects the visitor's gaze back toward the mountain (Engler 1999, 44).

A practical and sophisticated approach represents the tendencies of supermodernism. A couple of proposals seize the peculiar and adverse conditions of a dump and use it as a laboratory for new experiments. The Dutch team of Berend Strik and One Architecture Group created for Bavel an artificial moor in a pit at the top of the mound; it is intended to function as a catalyst that would apparently return the landscape within several years to its former state as a marsh. The addition of islands provides much-desired ground suitable for new luxury residences. Their proposal, "Artificial Archaic Landscape," uses the dump to link seventeenth-century Dutch landscape paintings and possible future suburban landscapes, overlaying classical images and contemporary situations. The Dutch artist Aernout Mik considers the Bavel site an extreme and displaced environment, and he dwells on its deceptive qualities. He appropriates the dump for genetically engineered animals. Mik's proposal, "Garbage Bar," is an expression of his belief in technology and the free choice of humans; he creates a distorted landscape complete with mutated cows, miniaturized to the size of dogs, in order to give visitors an illusion of a grand, far-reaching land (Driessen and Van Mierlo 1999, 106). He also places a half-buried glass pavilion with lavatories in the basement for convenience and personal illumination.

The ecological artists Avital Geva and Carsten Höller use the sites as catalysts to address and solve the problem of garbage disposal. They propose to build education and research facilities that will focus on waste reduction and technological innovation. Geva, an Israeli, proposes "A School for Garbage" at Hiriya. In working greenhouses fed by treated wastewater, Geva envisions scientists, experts, businesspeople, and students experimenting with agriculture, garbage, plants, and fish and teaching others about the practical, social, and philosophical aspects of garbage. Höller, who works in Cologne, resolved not to submit a design but instead to facilitate an intellectual discussion on the future of the dump. In his proposal, "Possibilities," the Breda landfill serves as a platform for artists, designers, scientists, philosophers, and engineers to exchange ideas and evoke possibilities of change. He shows possible models for buildings over the dump where this discussion can take place.

Finally, the architect team of Iris Horowitz and Morian Minuchin and the artist Micha Ulman each propose to prolong Hiriya's waste function in the form of a recycling plant. Horowitz and Menuchin turn the site into a recycling factory that is also open for the public. They transform Hiriya from a burial site for material culture to a place where the end and the beginning of consumer products meet. Micha Ulman would like to construct a huge sorting and recycling plant at the core of the mountain, after its content has been burnt. People will cooperate and sort the garbage in their private residences. The plant will be covered with earth and constitute an artificial nature preserve with openings for peering into the plant.

Committed to social responsibility and sustainability, the last group of exhibition proposals demonstrates the merger of practicality and economics that Van Dooren envisions.[17]

The dump, the epitome of a marginal landscape, has a special place in the social, literary, poetic, and artistic imagination: it represents a place where realities and fears are buried along with fantasies and myths. Outside these realms, the dump has no glamour or sex appeal, marking nothing but the mundane. It is meant to solve the problem of garbage, to doom it to oblivion. Yet the dump has become conspicuous, threatening, and hard to ignore. As the greatest earthwork monument of our times and a cogent symbol of our consumer culture, it is an exceptional terrain for social and design or art inquiries, a place where our rejected and silenced cultural values are awakened, provoked, and interrogated. It is a place where cultural aesthetics and disciplinary conventions are contained and affirmed, questioned and transgressed, and where latent social and economic opportunities await their turn.

A dump can be held as a mirror to our culture. It also has a unique capacity to transform social consciousness and condition. Only those who recognize the quality of landfills to provide us "an aesthetic distance from which to know ourselves" (Stegner 1962, 34) and a place from which to "imagine, find, or make the ways back to all of us," (Ammons 1993, 24) can reclaim the true spirit of these places. And only those who realize the dialectics of garbage can seize its potential to empower, to turn things around. For . . .

> where but in the very asshole of comedown is
> redemption: as where but brought low, where
>
> but in the grief of failure, loss, error do we
> discern the savage afflictions that turn us around.
>
> (Ammons 1993, 21)

CHAPTER 4

Waste Recycling / Reuse Institutions
Places of Material Transactions
and Resource Parks

. . . a monstrous surrounding of
gathering—the putrid, the castoff, the used,

the mucked up—all arriving for final assessment,
for the toting up in tonnage, the separations

of wet and dry, returnable and gone for good:
the sanctifications, the burn-throughs, ash free

merely a permanent twang of light, a dwelling
music, remaining: how to be blessed are mechanisms,

procedures that carry such changes!

A. R. Ammons, *Garbage* (1993)

To reinvent rather than merely mechanically apply recycling, it is necessary to rethink the relationships between the material world and culture and the "places of waste" where rejected goods can be circulated.

Polemical Matters
The garbage machine

One year after the first Earth Day, the town of Franklin, Ohio, held an impressive ceremony for its "garbage machine." The *Franklin Chronicle* boasted in its headline of the "World's First Garbage / Trash Recycling Plant," adding in the accompanying article, "Every resident of this comparatively small industrial city simply by taking out the garbage, now practices total recycling of household refuse" (Kelly 1973, 101). The plant opening on August 11, 1971, was treated, "as a truly all-American day," writes Katie Kelly in *Garbage* (100). After the flag raising, national anthem, a performance by the high school band, and welcoming remarks, thousands of Franklinites trooped through the big facility for Public Inspection and uncovered a plaque that read "Dedicated to the citizens of this small community who had the foresight and courage to save the purity of the land entrusted by God" (101). This pilot project of the Black Clawson Company recovered glass, sand, paper, and metals and shredded the residue into small particles, whisking it pneumatically into a devouring 1500° F burner. The mechanized plant was heralded as the ultimate solution to the garbage and virgin resource crises, and recycling was heralded as the country's salvation.

The proud Franklinites (and other avid recyclers of the 1970s) probably did not know of a much earlier garbage machine that undercut their claims of novelty. Almost a century earlier, in 1884, an article in *Metal Worker* featured "a strange machine" installed at New York's Jackson Street wharf. It was designed to "turn street sweepings and house refuse of all sorts into money" and was described as "a vast rag and bone picker of many Italian-power, working by steam" (quoted in Strasser 1999, 130–31) (fig. 4.1). The machine separated materials by shaking them, screening them, and putting them in water to float or sink. Italian immigrants finished the job, picking out the usable materials off the conveyor belt.

The perception of garbage machines as a panacea for the solid waste problem reflected America's long-standing trust in the mechanized world. What was new in 1971, however, was the broader public acceptance of recycling. Around the same time the Black Clawson plant began operation, environmentally motivated groups in hundreds of small American towns opened more modest recycling centers. Over the past three decades, large and small American cities have witnessed the birth of waste "redemption" institutions of all sorts and names—recycling centers, recycleries, material recovery facilities, resource recovery plants, material redemption sites, community material resource parks, drop-off sites, buy-back centers, and refund centers. The pervasiveness of recycling and its host places, whether public or private, for profit or for environmental gain, demonstrates a new sense of the connection between consumption and material waste, and it promotes new rituals of social-economic exchange with a moral-environmental tinge. The places we have created to aid reuse and recycling serve vital and potent roles in the contemporary landscape, and not for waste management purposes alone. Complex political, social, ideological, and aesthetic realities have been produced and sustained by these compelling places of material transaction.[1]

Discourse about recycling and recycling centers can easily drift into self-congratulation. After all, popular recycling, often a mere act of placing one's used matter in specified containers for removal and processing elsewhere, is an act of environmental consciousness, of participation in a common global goal, the "right thing" to do. Such complacency vanishes, however, when adversaries attack the "business" of recycling and accuse the faithful recyclers of acting mindlessly, duped by the new, powerful recycling industry into supporting the established consumption-based economy. This dispute raises important political, economic, environmental, and ethical questions, which remain unresolved and harshly contested.

The politics of recycling

In the United States, recycling, the great moral crusade of the late twentieth century, has become a normal, almost ubiquitous activity.[2] Even as growing numbers of people act as if they think that recycling is worthwhile, environmentalists and economists are turning more doubtful. By now, everyone knows that there is no landfill crisis and that recycling is not saving us money. In seeking the question to which recycling is the answer, we find ourselves exploring the connection between recycling and larger environmental issues. For while recycling is not always profitable in the short term, claims the solid waste researcher Frank Ackerman, it is nonetheless a response to a long-term environmental problem that cannot be reduced to narrowly economic terms (Ackerman 1997).

Those engaged in today's fierce debate over the recycling agenda do not simply fall on one or the other side of the traditional line dividing anticonsumerism environmentalists and corporations. Recycling's avid proponents believe that it conserves natural resources, prevents pollution caused by the processing of either raw material or garbage, saves energy and landfill space, and, not least, engenders a sense of community involvement and responsibility. Their opponents dismiss these claims and attack recycling on economic, environmental, and even moral grounds. This position was encapsulated in a controversial 1996 *New York Times Magazine* cover story titled "Recycling Is Garbage," in which John Tierney (1996) argues that most recycling efforts are economically unsound and of questionable environmental value. Recycling requires a complex industrial process more costly than landfilling, creates additional pollution, distorts free-market objectives and efficiency, and diverts federal funding that could be used to subsidize more worthy environmental causes.

For the most part and from the point of view of the individual, say critics, recycling relies on a narrow and partial notion of environmental responsibility. It is popularized and legislatively enforced because it per-petuates the usual consumptive habits, supporting the interests of the same corporate industrial system that created the waste problem in the first place, as well as the new parasitic recycling industry. Municipalities, too, are trapped in the larger economic system; after they put in place large-scale, wasteful programs and facilities, recyclables often must be dumped because of lack of demand or their low market value. In short, these critics' opinion is that recycling is not only a waste of energy, money, and resources but also a means of deceiving the public. They in-stead favor programs that conserve at the resource level and renew bio-mass material, as well as make the manufacturer responsible for reduc-ing packaging and devising new sustainable materials and products.

What began as a grassroots counterculture effort has been institu-tionalized, politicized, used as a public relations tool, and even com-modified and absorbed into consumer culture. The momentum created by 1960s environmentalists and hippies and their collectives and al-ternative businesses—food co-ops and recycling centers—was, by the late 1980s, propelling forward large waste management corporations. Though there initially were more than ten thousand (some say tens of thousands) independent companies, recycling in all its functions—col-lection, landfills, recycling and transfer stations, and high-tech incin-erators—is now managed primarily by a handful of giant companies. This new multibillion-dollar industry instituted the National Recycling Coalition to advance its political agenda—to end virgin material subsi-dies, finance recycling businesses, promote zero waste, and so on—on a not-very-receptive Capitol Hill. The industry also sponsors a yearly America Recycles Day on November 15, an event endorsed by President Bill Clinton. It first took place in 1997, with the goal of promoting re-cycled product sales and pressuring Congress to favor pro-recycling leg-islation. About 3 million people across America participated in thou-sands of events: parades, raffles, public sales and displays of recycled products, and sign-ups to buy recycled goods. A few recycling prizes were awarded, including one from Disney World given to a youth re-cycling program. The Mall of America in Bloomington, Minnesota, the ultimate symbol of America's consumerism, allocates a special store to promote recycling and takes an active role in America Recycles Day. The store labors to show off its "green" virtues, exhibiting information on the recycling bins and collection system within the mall, and, of course, selling "postconsumer" products year-round.

Regardless of the criticism and corporate clout, recycling has become a popular cultural ethos matching, if not replacing, the disposability ethos of the not-too-distant past. The new regime of social behavior is enacted with an aura of environmental stewardship. Viewed as environmentally positive, the burgeoning recycling industries enjoy loose regulations, subjecting nearby communities to pollution, noise, and smell (fig. 4.2). Probing the multifaceted political nature of recycling in a study focusing on Fairfax, Canada, the social scientist Eric Darier (1996) claims that it has become a legitimate topic of discourse and a modern form of power. Discourse about recycling unfolds during public hearings on waste management policies and facility siting, when various interest groups articulate their stances and expose the power relations that prevail in the system. According to Darier, recycling as a form of power is both a normalizing instrument and a means of resistance. As a subject of normalization, it is used by government or corporations to encourage uniform behavior that advances a specific aspect of the core agenda. The government wants to reduce its own spending on waste management and environmental concerns while facilitating the general growth of industry in these areas. It therefore forges campaigns to educate and inform individuals, using them as key instruments in reshaping people's social conduct. The tactics used, however, are not intended to teach citizens critical skills (for their own or the environment's good), claims Darier, but to train them to act in a particular social mode.

According to Darier, conventional recycling, in being normalized, has also encountered two types of resistance. One type arises from people working against the institutional status quo who begin to respond to issues related to garbage in a radically different ethical way. For example, some groups or individuals create local programs, such as composting farms and building materials reuse yards, and promote the active reuse of "waste" material by the entire population. This type of resistance has shaped a new social movement, involving subversive groups and individual action. As a consequence, however, the local agencies and corporations in charge also resist relinquishing their power to the masses by preventing the decentralization of waste management. Darier concludes that today's recycling can be considered an expression of a new group and individual identity, of a "conserver society," a "green self" or a "postconsumer self"; it is a form of resistance to consumer culture that can be coupled with a social movement (1996, 81).

The psychology and sociality of recycling

Recycling behavior illuminates how we think about the environment and about the consequences of our actions. The popular motivation for recycling is undoubtedly complex. Social scientists eager to unravel the new middle-class recycling habits conclude that recycling is spurred by some combination of extrinsic and intrinsic goals—monetary incentives, utilitarian and personal gains, and ideological and political motives. Early studies in the 1960s found that financial inducements (e.g., refunds for cans, payment by weight) were a major factor in recycling habits. Studies in the 1990s, however, echo Darier's assertion, citing motives related to personal satisfaction, such as self-sufficiency, making a difference, and being part of the community, as more critical (Oskamp et al. 1991). Some researchers stress the moral aspect and claim that like other actions taken in favor of the environment, recycling may be commonly understood as a middle-class altruistic behavior guided by social and personal norms (Hopper and Nielsen 1991).

Although recyclers may be interested not in fundamentally changing the system but rather in making the environment a bit cleaner and alleviating their own guilt, they use recycling to articulate a new social and self identity. (Sometimes the same people who devotedly separate their recyclable and advocate recycling object to the siting of recycling plants in or close to their neighborhoods.) Frank Ackerman summarizes the reasons for the broad support for recycling and concludes, "something is wrong; garbage is involved; experts have failed in their attempts to diagnose the problem; cooperation with the forces of the market is inescapable, but cannot be the whole story" (1997, 20). In addition, he claims that although people know that reducing or preventing waste is more effective, they find recycling easier.

The new green self announces itself in more than the mere act of separating recyclables in preparation for a weekly pickup. People rekindle the social functions that are latent in shopping for and exchanging material and objects and apply them to transactions involving used commodities. Although recycling and reuse are inversely related to consumption, they also have some similarities to it; both help forge group identity. Daniel Boorstin (1973) has observed and written on the phenomena of "consumptive communities" in the early twentieth century. In his incisive analysis of contemporary consumer culture and the social and communal patterns that arise in places of consumption, the geographer Rob Shields makes a case that consumption itself "has become a communal activity, even a form of group solidarity" (1992, 110). Today's consumer, he contends, is an informed social actor who consumes the symbolic value and image of the commodity and shopping environment while actively participating in the creation of a public sphere. The mall environment hosts community events, fairs, and holiday celebrations while providing its crowd with a space to actualize different group identities. The spaces of consumption therefore claim some of the aura of the market square and earlier street rituals, and consumption is "a form of social exchange through which community, influence and micropowers (in the Foucauldian sense) are actualized" (1992, 99).

Shields's understanding of consumption can easily be extended to recycling and its assigned spaces. Recycling and reuse, though ascribed a different symbolism, also help create communal-social settings. The same middle-class consumer who buys goods at the mall also partakes in the new recycling and reuse environments, where consumption and retrieval overlap. Individuals join in a convivial social interaction and exercise new rituals of community and membership as they accept a grand ideology concerning community, locality, and environment. People use recycling as a means to find new places for gathering and exchange, places that reproduce the old models of institutions dealing in second-hand goods and enable individuals to partake in the new environmental ethic.

Recycled language

Recycling has become popular so recently that it has not yet acquired a decent vocabulary. The four related terms found in dictionaries are dull and ambiguous: recycling, reuse, and adaptive reuse, or secondary use. Regrettably, all are often lumped together under the term "recycling," which blurs their distinctions (and helps keep the rhetoric politically correct). In fact, each term has a specific meaning, indicating different levels of energy invested and alterations made as refuse is transformed into reuse, as well as different implications for how the material in question was initially produced.

"Reuse" denotes the continuous use of some object, matter, or space with little energy invested or physical change, which can be done by individuals with minimal skills (e.g., refilling cleansed glass bottles, writing notes on scrap paper). It assumes that the material goods in question are durable, and it therefore requires the adjustment of the entire current production system. "Recycling" means different things to different groups. For a garbage producer, it entails placing the waste matter in a designated container for removal by the recycling truck. For the consumer, recycling means the production of new postconsumer goods. In professional terms, it entails the physical diversion of garbage from the waste stream intended for landfilling or incineration and its reprocessing into new material resources. Professionals often refer to recycling as "source separation" (originally called "primary separation," in the 1890s).

A great amount of new energy is necessary to recycle this "waste" into a new product, while the energy that was invested during its original production is lost (Pawley 1975; Wilson 1993). For example, the energy required to endow tires and plastics with structure and their great strength is wasted when tires become rubber pavers and plastic milk jugs turn into plastic lumber. Recycling is industry-based and cannot be done by individuals. It creates new energy-consumptive operations and calls for no change in patterns of consumption. "Adaptive reuse" and "secondary use," used as synonyms, denote another, more creative and imaginative method that has been known for years. Adaptive reuse gives a new application or function to the same object, space, or matter, which requires only slight alteration. The term has been used extensively by Kevin Lynch to denote a new use for an old space or area, and its meaning is close to the idea of "reclamation" or "rehabilitation." While recycling maintains our dependence on the existing economic system centering on industrial mass production, reuse and adaptive reuse contribute to greater personal independence and self-sufficiency but often in ways that do not support established industry or satisfy popular taste.

Dictionaries have little to say about the key words in this field. "Recycle" and "reuse" bear the prefix re-, meaning "again" or "back." When re- joins cycle, it doubles the meaning of the root word, which already means a recurring pattern of events or phenomena. When the word "recycle" was coined in the 1920s it referred only to the reuse of materials in an industrial process (for example, returning gasoline produced as a by-product back into the system). It was first applied to discarded waste material in the 1960s (OED, 1989). It emphasizes the changes and treatments necessary to regain a material of usable form, an adaptation for human use, or the original condition. "Recycler" and "recyclable" were added only in the 1970s. In dictionaries, "reuse" implies returning to its previous use something that has become useless; "adaptive reuse" or "secondary use" entails the adoption of a new and different function for an unused object or space.

The prefixes *un-*, *up-*, and *down-* have also been put to use. Lynch introduced the term "*un*cycling" to suggest a need to refrain from recycling everything, and other scholars used "*up*cycling" and "*down*cycling" to suggest the qualitative downgrading or upgrading of material during the recycling process. Materials retrieved from the waste stream are called "scavenge" or "salvage" material (the latter implies an act of redemption). The products of recycling are called "recycled goods" or "postconsumer goods" in distinction to "virgin materials," which are extracted directly from nature. While the term "virgin" clearly implies something free of impurity, fresh, and unspoiled, "recycled" suggests a second version and thus a degree of contamination

The idea of pollution is but one of three underlying constructs inherent in the idea of recycling.

The ideology of recycling

From the consumer's perspective, recycling is a process that re-creates entirely new goods while eliminating the threats and stigma attached to used goods. Recycling is founded on the premise that an item's condition as waste is not inherent or absolute; rather, waste can acquire a new, value-added status. The social scientist Michael Thompson posits that for a waste object to transcend its waste status and acquire value, three transformations are necessary: its life span must increase, it must emerge from obscurity or anonymity, and its polluting properties must be removed (1979, 26). Recycling, more than any other form of reuse, intervenes to extend durability, ensure an identity, and erase defilement.

Recycling and pollution. Used and reused things seem polluted on both sensory and symbolic levels. Used things smell and look unclean, gray, and a little greasy. We object particularly to traces of other humans because we fear contamination. These traces exert a kind of horrific magical power on us, notes Terence McLaughlin in *Dirt:* "if we happened to touch them or even smell them . . . [we fear] that these unknown people who have been in place before us may somehow infect us with their own diseases and shortcomings" (1988, 5). Another symbolic contamination draws on the sexual metaphor of exclusive possession: used things have a nonfresh, nonvirgin quality (Lynch 1990, 19). Recycling eliminates the sensory qualities of smell, griminess, and filth, as well as the marks of past use. As material goes through the production process again it is purified, thereby satisfying the contemporary postconsumer.

Recycling and identity. Waste is obscure and anonymous, often an indiscriminate mix of items. Visually, it is chaotic, lacking pattern and identity. As recyclables are separated and put into distinct containers with like materials, they exhibit a kind of specificity. Moreover, the process of recycling turns old products back into raw material that is recast anew, often ground and melted before taking on a new form and identity (which often ignore the object's original identity and history).

Recycling and duration. Garbage implies the practical and cultural death of matter, the end of the object's life trajectory or flow. Different materials have different trajectories: glass may last for 100,000 years; organic waste may wither within days; humans expect to live less than 100 years. Culture is the system that organizes such flows. In contemporary culture, which develops and loses desires quickly and assumes that objects should be enjoyed briefly and then thrown away for newer and improved replacements, recycling has been given the authority to use technology to recover objects, to extend their useful and meaningful life.

The notions of duration and flow necessarily bring us back to the idea of entropy, increasing disorder in the universe. Recycling and entropy are, in a way, contradictory concepts. As noted in chapter 1, the second law of thermodynamics implies that everything in the universe begins with structure and energy and is moving toward random chaos and decay, and therefore waste. Although there is no way to escape this physical rule, our primal instinct is to oppose it and slow down the process of decline. Recycling, one such response, works as a rational tool to slow temporarily or reverse the inevitable loss, and therefore to lessen human existential anxiety and promote continuity and order.

In his prescriptions for "good wasting," Lynch tells us that our principal policy should be to keep resources circulating without irretrievable loss and that we should not attempt to shorten or extend an object's life. We need "to fine-tune the duration of things" (1990, 191). For Robert Smithson, recycling garbage as an attempt to reverse or slow down entropy is "hopeless and pitiful"; waste represents an entropic condition that should be accepted and need not be reversed (1996, 309). He approvingly quotes the renowned economist Nicholas Georgescu-Roegen in his 1971 book *The Entropy Law and the Economic Process:* "I do not think things go in cycles. I think things just change from one situation to the next, there's really no return" (quoted in Smithson 1996, 309). Rather than attempting to reverse waste, Smithson strove to shape it into something new, without returning it to an original form or something that would be considered useful.

The discussion on the need and means to counter entropy is ongoing. The solutions that society has put forward, all economically, politically, socially, and ideologically bound, have shaped myriad places of material exchange.

Sites of Material Exchange

Early recycling places

The reuse of waste is a practice as old as humanity. Specialized private spaces (at one's home, yard, field, alley, and street) have always been allocated for storage, repair, and exchange (see chapter 2). Municipal and industrial reuse and recycling places, originally built to relieve the city of its offal and excrement and turn that refuse to profit, generally have been established only within the last two hundred years, although there are a number of exceptions from antiquity. Ancient Athens had an effective composting plant, and Knossos, the capital of Crete, had a composting pit for organic waste in 1500 B.C.E. (Kelly 1973, 16).

At the outset of the nineteenth century, France had already acquired a name for its highly refined *poudrette,* a fertilizer and fuel made of dried excrement that was processed at Montfaucon, outside Paris. In the American industrial arena, reuse and recycling operations began showing up near large urban slaughterhouses; they turned leftover animal parts into glue, soap, fertilizer, hog and chicken feed, combs, and more. Plants in Chicago and New York produced everything from cosmetics to grease to explosives. By the 1850s the stench and offal in New York's streets were such a nuisance that the city had to take control of matters. It encouraged the establishment of private offal industries; leased them in far-off locations on South Brother Island, at the mouth of Jamaica Bay, and later on Barren Island; and contracted them to remove and process the city's offscourings (Miller 2000, 43). In *Fat of the Land,* a thorough account of New York's garbage saga over the past two centuries, Benjamin Miller describes the result: "Within a decade, [Barren Island] had the largest concentration of offal industries in the world, and the white sand, sedge, and cedar were transformed into loading docks and smoke stacks and factory buildings and workers' shacks. The island that the Indians had called Equindito (Broken Lands— . . .), the English settlers, through a natural linguistic evolution called Barren. . . . [T]he only vessels that approached it were scrows filled with offal" (2000, 44).

In addition to the slaughterhouses and pig farms, noxious industries soon located near the new offal recycling plants. Other, less offensive refuse, such as ashes, was especially sought-after for reuse at cement plants. Horse manure from the streets was sent to fertilizer manufacturers, and organic garbage was fed to livestock. Strasser's historical review reveals that in the last third of the nineteenth century, as technologies that made possible the recycling of bones, rubber, and steel were invented, new private manufacturing emerged that relied on city supplies (1999, 102–5). "Reduction plants" using a new process began "reducing" municipal waste with new machines and chemicals to produced new by-products, such as pharmaceuticals, explosives, lubricating oils, and cosmetics (134–35).

The first time that the concept of recovering value from municipal waste was put to work at a city manual rubbish sorting plant was in 1896, during the tenure of the renowned sanitary engineer Colonel George E. Waring, who acted as New York City's street cleaning commissioner between 1895 and 1898 and who coined the phrase "Waste to Wealth." In order to make collection easier and increase profit, he imposed a system of "primary separation" that required householders to store wet garbage, rubbish, and ashes in separate containers. He assigned forty police officers to explain the new rules and enforce compliance (Melosi 1981, 70; Rathje and Murphy 1992, 175). Many of Waring's initiatives terminated soon after his death in 1898 but were restarted in 1902. During the first two decades of the new century, most cities with populations greater than 25,000 required their citizens to separate organics and ashes and deliver them to their respective processing sites (Strasser 1999, 129).

Garbage incinerators, also called "destructors" or "cremators," were patented in Britain but soon came onto the American scene, offering complete elimination of garbage with no need for separation. According to Strasser, the first was built in 1885 on Governors Island, in New York Bay. An early model opened for public inspection was triumphantly displayed in the sanitary White City of Chicago's Columbian Exposition (1999, 131). Incineration was promoted as the ultimate method for cleanliness and convenience. Incinerators presented serious competition to recycling, although for some time immigrants were hired to pick through municipal trash and recover marketable material from the conveyor belts feeding the ovens. The facilities were located in marginal, grimy industrial urban areas and near dumps. The small facility was dominated by a smoke stack; garbage entered via a ramp on one side and exited as smoke and as ash that was loaded on trucks and dumped or used as a fertilizer and fill cover. Incineration became popular in the decades between 1890s and 1930s, despite attacks by those who believed in the value of reusing material and considered the burning of waste a sinful extravagance.[3]

After the start of the new century, as people were gradually caught up in the growing consumer economy and throwaway ethic, places devoted to reuse disappeared almost completely from the private, public, and industrial domains. Consequently, municipalities built more incinerators, used garbage for land reclamation (landfilling), and allocated more dumps for the remaining waste. The two world wars and the depression were exceptional periods, when campaigns organized by the government and civilian organizations made recycling a mark of patriotism. During World War I the government encouraged household thrift and enforced industrial frugality. Propaganda urged citizens to conserve food and clothing, and other materials—including paper, wool, rubber, and metals—were regularly collected. The U.S. Waste Reclamation Service, created in 1917 as a section of War Industries Board (later joining the Department of Commerce), initiated programs in selected communities, but the crisis dissipated within months. During World War II the efforts were more widespread and longer lasting, and grassroots conservation movements mobilized scrap drives. Large-scale campaigns to collect scrap metals, grease, paper, rags, fat, and rubber joined other government programs such as rationing and production restrictions, and they produced huge quantities of materials (fig. 4.3). Sorted waste piled up in vacant lots, schoolyards, and junkyards and became a familiar sight. Junkmen were contracted by the government to collect and transfer the materials. During the wars, mending, reusing, and donating material were moral acts in which everyone, especially schoolchildren and women, participated (Hoy and Robinson 1979, 15; Strasser 1999, 19). People were even asked to tell on their neighbors who hid strategic war material such as iron. Waste was then called "urban ore."

Soon after the war, disposability rose to record levels. Cities that had previously required citizens to separate refuse rescinded those regulations. Waste changed significantly as plastic materials and paper began dominating the waste stream. Organic waste decreased as garbage disposals installed in most new kitchen sinks began turning food into sewage, discharging it directly into the public drains. A few composting operations started after World War II but failed soon after opening. Land-fills came to rule the disposal landscape.

The new recycling landscape

A new wave of recycling industries emerged in the late 1960s, following a few decades of little to no recycling activity. The first recycling places were "buy-back" facilities, which opened in response to the high costs of dwindling virgin resources, especially aluminum and paper. Cans and bottles were refunded for five cents each; together with newspapers, they were collected and sold to companies that began experimenting with re-cycled material around 1967. The early promoters of recycling were members of the counterculture, hippies and environmentalists. Around five hundred recycling locations were opened in twenty-one states be-tween 1970 and 1971; many more soon sprang up around the country, operated by both the private sector and activist groups (Kelly 1973, 123).

Some states passed "bottle bills" requiring buyers to pay a deposit on bottles and cans; this step was designed to solve the urgent aesthetic problem of litter tossed along roadsides as well as to replace expensive virgin aluminum.[4] It also helped affirm the idea of recycling and spurred the creation of many refund centers (Rathje and Murphy 1992, 201). These were usually built as temporary sheds in industrial edge locations. By 1979 an estimated 5,000 citizen-run, municipal, and small commer-cial collection centers were in place (Hoy and Robinson 1979, 6). People were extremely willing to participate, but the markets and consumer de-mand for recyclable material did not emerge until well into the 1980s. Despite the enthusiasm of the involved activists, many of these sites soon closed. There was more material available than anyone wanted to buy at a reasonable price, and the uncertainty of the new system halted the development of recycling. In the meantime, large waste companies began dominating the business of collection and processing and re-placed the small, primitive centers with bigger, mechanized plants. In the 1980s, many new community programs—drop-off sites, curbside collection, materials recovery facilities—and federal and state laws and incentives provided the needed push.

Gradually, supermarkets opened their own refund stations. Conve-nient drop-off sites lined with containers showed up in supermarket parking lots. The "ATMs for recyclers," self-serve automatic machines that whisk away, sort, and tally a variety of cans and bottles using an electronic eye and sensors, were added beginning in 1991. One of the first models, the Recycling Eagle, was patented by the Californian inventor Steve Aslanidis (with David Vosburn, a nuclear engineer) and placed outside major food stores. The easy-to-use, friendly, yet strict machine speaks to its users, homeless and shoppers alike, giving them instructions and scolding those trying to cheat (Greene 1993, 20). Yet the ultimate in convenience was curbside recycling programs, which began in the late 1980s, primarily in suburban, single-home neighborhoods. Curbside recycling spared householders the chore of having to deliver the material and had two other consequences: it forced many small drop-off redemption sites to close and helped waste companies monopolize recycling. It also changed the residential street scene. On certain days, the sidewalks now offer a brief display of containers of colorful sorted materials. At last, recycling took hold and became a presence in suburban neighborhoods.

In the 1990s growing social support for the recycling markets and the new convenient ways to recycle dramatically increased the participation rate in recycling programs. In many cities and small towns, recycling changed from voluntary to mandatory. After lawmakers banned yard waste from some municipal landfills, composting facilities in large cities mushroomed and helped divert 20 to 30 percent of the municipal solid waste stream.[5] Popular middle-class recycling and domestic waste management magazines completed the picture. *Garbage* magazine educated its green audience about practical saving, recycling matters, and options that would help them "save the world." *BioCycle* magazine targeted gardeners and promoted self-sustaining organic gardening. These publications recast the kitchen and the yard as spaces for sorting, processing, and reusing waste and promoted a new aesthetic, which began to loosen the grip of the tyranny of cleanliness and embrace a measured amount of pleasant untidiness. For millions recycling became a mainstream activity and a fact of life.

Since the 1970s the landscape of municipal solid waste facilities has expanded gradually to include, in addition to landfills, a range of manual and mechanized recycling, transfer, and waste-to-energy plants. Small, manual centers have been located in both large and small towns. Lined with large containers and tanks for already-sorted recyclables, they service trucks from curbside recycling programs and cater to individuals, especially the homeless who come seeking cash in exchange for items they have collected. Sometimes these places perform the initial process-ing—washing, shredding, and baling the material. Large, mechanized recycling centers, also known as material recovery facilities (MRF), have mostly been located near operating landfills and in heavy industrial ar-eas further away from the city center. They receive mixed garbage and separate, process, and transfer the sorted materials for further produc-tion. Their massive boxlike warehouses contain giant mazes of assem-bly-line machines connected by branching conveyor belts—screens, shakers, shredders, hammers, rotating drums, magnets, compactors, balers, and chutes (fig. 4.4). Garbage trucks feed a giant tipping floor and a trommel on one side, and transport trucks receive their cargo from chutes at load-out structures on the other side. In between are manual picking stations for undesired or salvageable items.

Transfer stations, built to facilitate the long haul of garbage to newly opened distant landfills, also joined the garbage scene in the 1990s. They consist of gigantic spaces with a tipping floor, a large concrete trench for garbage compaction, and a leachate purification system. Construction is usually plain and light (though impressive structures have recently be-gun showing up). Large facilities in big cities often combine the two functions of recycling and transfer in a new hybrid called a "trans-cyclery." The transcyclery sorts and diverts some materials before ship-ping the rest to its burial. Scales to weigh the waste and recycling trucks are essential elements of all these places.

Meanwhile, owing in part to the energy crisis of the 1970s, a relatively small success was garnered by another "recycling" institution. The waste-to-energy (WTE) facility, also called the resource recovery facility, has attempted to ride the wave of recycling's popularity. Though it claims to make waste useful, it is in fact a version of the notorious in-cinerator.[6] WTE facilities came under attack from environmentalists, whose goal of saving virgin resources they threatened. This type of fa-cility failed to win the hoped-for popularity because of its rejection by environmentalists, fears about air pollution, its complicated mainte-nance, and its costly operation. But with better pollution-monitoring systems it has recently again drawn interest in some highly populated parts of the world.

While private garbage and recycling machines were being perfected and becoming efficient industrial facilities, a new industrial concept that offers a completely different way to deal with waste was also evolving. The eco-industrial park is a response to the wastefulness inherent in the current system of recycling. This type of industrial development challenges traditional forms by creating what might be considered a wasteless environment, omitting waste storage and processing altogether. It is based on a closed loop or interconnected system in which one industry's waste output is another's raw material, making a chain of use and reuse. Researchers in the new field of ecological industry have been perfecting this concept since 1995, as they seek a greater harmony of ecological infrastructure, or the so-called ecostructure, with the Earth's unique biogeochemical processes. They aim to tune the industrial system to natural cycles and flow of matter.[7]

Traditional institutions dealing in the secondhand

As varied industrial waste recycling institutions were being shaped, other places that emphasize the social aspects of used items also developed. These endeavor to reclaim the lost human contact with used material culture, the social ambience associated with secondhand exchange, and the social benefits waste once offered the underclass. Contemporary recycling and reuse places expand on the intentions of traditional institutions dealing in secondhand goods, which for many years were instrumental to the circulation of waste and the dynamic processes of reuse, feeding off the affluence of the city rich and serving the poor (Strasser 1999).

Trading in reusables provided a living for thousands of sellers, middlemen, and dealers and produced special public, semipublic, and private places of vital socioeconomic exchange. Until the 1960s these places were associated primarily with the lower class, but as waste turned into a resource and used items became antiques, shopping for secondhand goods became more acceptable and fashionable among the middle and even upper classes.

Located not far from the town market and confined to a small, specialized area, the flea market is one of the oldest recognized reuse institutions. It originated in the 1870s along the northern wall of Paris and took its name from its vermin-ridden merchandise. While the town market was a central hub of commerce and festivity for the well-to-do, the flea market served the poor on the margins of town. Still, it had its own gaiety. The flea market was a productive hub of trade and traffic in used goods; there, ragpickers brought the items they collected at city dumps and on city streets. This system of used-goods redistribution was subjected to regulation and suppression on both political and moral grounds. It became popular in mid-twentieth-century Europe; only since the late 1960s, in a slightly different form, has it gained favor in the United States as well. Until then, American culture with its strict patterns of commercial exchange did not lend itself well to the flea market or to open-air markets in general, except for selling farm produce. In the middle of the twentieth century only a few open-air markets existed; used materials were dealt with through other institutions.

The thrift institution also played a central role in circulating used goods. Unlike the flea market, whose secular atmosphere reflected an economic focus, thrift stores were patronizing and imbued with moral and religious tone. Susan Strasser, who devotes a large part of her book to the major thrift organizations—the Salvation Army, Goodwill Industries, and the Society of St. Vincent de Paul—argues that they "both benefited from and contributed to the identification of recycling and reuse with poverty" (Strasser 1999, 141). These charitable organizations, which came into existence around the turn of the twentieth century, offered the middle class a righteous outlet for unwanted things and the poor "jobs, spiritual salvation, and a chance to be consumers" (140).

The Salvation Army began operating in New York City in 1878 and employed the poor in mending and repairing articles, which were then sold at low prices, while engaging in social and spiritual activities aimed to uplift souls.[8] "Industrial homes" housed the workers and workshops, and people pushing handcarts brought in the donations collected. The 1903 Salvation Army's New England annual report announced, "We confidently believe and are seeking to demonstrate that the want of our cities can be met from its waste!" (quoted in Strasser 1999, 144). Similarly, Goodwill Industries, which originated in Boston's Morgan Memorial Cooperative Industries in 1895, was part of a broader program that operated social and religious programs. It catered to people of more diverse religious backgrounds and took a more liberal stance. The "opportunity bag," a large bag placed in middle class homes to ensure donations, became its symbol. While the Salvation Army and Goodwill emphasized employment opportunities for the poor, the St. Vincent Society, which became effective in large cities in the 1910s, stressed charity. The thrift organizations took over run-down buildings in declining areas of downtown and slowly modernized them. Their early repair workshop operations were changed into retail-style department stores, where goods were displayed more attractively. In the 1920s, other thrift stores were opened by Junior Leagues and women's hospital auxiliaries, and the institutions spread in the 1950s. The middle class has always supported the charities but rarely buys from them (Strasser 1999, 280).

The junkyard has also long dealt with salvage material. Strasser tells us that unlike the thrift institutions, junk dealers did not benefit from the benevolent feelings and support of the public; indeed, they had to compete with the charitable organizations, which enjoyed nonprofit status and could undersell them. Moreover, the public had always stigmatized and mistrusted junkmen and their enterprises, considering them manipulative profit makers and their yard a blighted dump. Yet junkmen were key to the process of collecting material, especially during the world wars, when the government contracted with them to gather and deliver the fruits of scrap drives. Junk stores and junkyards were mostly located in light industrial areas or on the fringes of downtown in both large cities and small towns. But recently they have been losing their stigma. In fact, each of the three institutions—flea market, thrift store, and junkyard— has since the 1970s become relatively respectable and popular, taking on new symbolic roles that reflect the influence of nouveau riche aspirations, new environmental sentiments, and countercultural communities.

The new secondhand exchange institutions

In the late 1960s flea markets began showing up in large American cities, frequented not only by the lower-class customers but also by middle-class opportunists and treasure hunters. The markets are composed of temporary open-air or enclosed booths arranged in rows on collapsible tables, sometimes at the rear of trucks or station wagons. Concession stands offering refreshments and portable speakers announcing sales and playing music make the flea market a carnival-like event. At the weekend flea market in the large parking lot of the Ashby Avenue subway station in Berkeley, California, hippies, refined ladies, and everyday merchants sell a range of merchandise from antiques to necessities; there are also political booths where petitions are signed. Many flea markets have become tourist attractions, offering an exotic spectacle of sights, foods, and sounds; they have the appeal of liminal carnivalesque places, where opposite categories of high and low, commerce and festivity commingle. As the flea market has been gentrified, the shopping mall has become more popular and communal.

Upper-class shoppers have also recently found an outlet for their avant-garde desires in antique stores and other fashionable establishments selling "vintage" clothing and jewelry. These stores or boutiques are new versions of the thrift store, without the connection to charity; they began gaining popularity in the 1980s in tourist and gentrified urban areas; they focus attention on the uniqueness of old, durable, handmade items as compared with the short-lived, uniform, and mass-produced goods found in department stores. SoHo in New York is one such district. Artists' studios and galleries have been forced out to make space for the upscale second-hand shops.

Some contemporary junkyards have been transformed into large, "clean," and pleasantly organized outdoor operations that look like giant garage sales. Organized into specialized sections of goods and materials separated by paths (like flowerbeds in a garden), they offer both unique and mundane items at affordable prices and enable reuse to take on broader social functions. Moreover, the junkyard has gained respectability among other environmentally responsible recycling institutions. Urban Ore is one such operation in Berkeley, California. It gets its merchandise primarily from residents delivering their unwanted possessions and from workers sorting through city waste. The owner, Dan Knapp, a former university professor with a Ph.D. in sociology, recognized that a great economic potential is buried in garbage and decided to put into action his belief in total recycling. He made himself into an expert by working at dumps with salvage material. In 1980 he rented a lot where he and two other garbage colleagues sold building materials retrieved from the Berkeley dump; they gradually expanded into two salvage yards and a swap shop, hiring more than twenty employees.

No "garbage" or "waste" words are ever applied to the operation. The first yard, called the Building Materials Exchange, displays salvaged doors, windows, bathtubs, and sinks; the second, known as the Discard Management Center, offers furniture, clothes, and kitchen appliances. The name Urban Ore alludes to the World War II scrap-saving campaigns and to the fundamental notion that the city's waste is the city's wealth. Knapp is not only a dealer and manager but also an activist who preaches on recycling at conferences and in publications (McDonald 1990). Like other purveyors of the secondhand, Urban Ore has become an economically profitable business and a favorite setting for social and economic exchange among a range of customers who cross class lines, from the needy to treasure hunters and antique dealers.

As the old institutions have transformed their image and character, new places for exchange have sprung up. Garage and yard sales have gained popularity, losing any associations they may have had with destitution or environmentalism. They serve as temporary neighborhood marketplaces, informal offshoots of free enterprise. In this communal, social sphere, located on private grounds, objects no longer desired by the seller move directly to the buyer's hands. The researcher Max Miller traces the origin of the yard sale to the old trader or hustler, who used his yard for trading used merchandise outside normal channels (1988, 55). Garage sales, however, are generally recognized as having started in the 1960s as a suburban pursuit. Middle-class women built on the idea of rummage sales, which were sponsored periodically by charitable groups and well-to-do women (who donated goods and ran the sales but did not buy there), as they began selling unused clothing and goods that crowded their closets and attics to make space for more new stuff.

These private sales—garage, yard, lawn, barn, porch, community, moving, and whatnot—acquired their names from their host spaces or occasions. The spacious garage, a storage space where unused items accumulate and dirty work and repairs take place, is a natural site for the event. The large paved driveway adjoining the garage acts as a private plaza, an interface between public and private that provides access and convenience. The front yard is, of course, another good location, highly visible from the street. Such sales are organized by individuals, several neighbors, or a whole neighborhood; they are advertised in classified ads in local newspapers and announced by hand-lettered cardboard signs attached to trees and telephone poles, and balloons and ribbons draw attention to the site. Although most garage sales are occasional, some individuals run them as a regular weekend business with items regularly on display. The highway yard sale is another, more structured version that is held along well-traveled roads, usually every weekend during tourist season. Attending these quasi-social sales has become a popular pastime. The multifaceted events blend practical household cleaning and fanciful treasure hunts, community recreation and individual profiteering (Herrmann and Soiffer 1984, 412). These private sales allow some people to exercise their bargain-hunting skills to profit in an informal economy. The small revenues gained remain untaxed, offering competition with charitable rummage sales and thrift stores. But because city neighborhoods are segregated by class, the needy have often difficulty participating in this popular activity, which is thus less effective in connecting supply and demand.

In their many forms, places supporting reuse make an ecological and economic sense. While linking material goods and people, supply and demand, they also create public hubs and community gathering centers.

Community material resource parks

New communal institutions have also emerged since the 1970s. They combine a grand narrative of the global mission of protecting the environment with local settings that actualize new community and individual identity. The sociality of traditional reuse institutions has also taken shape at these newly created recycling places. The early small drop-off sites at various spots in the city spontaneously began to function as casual meeting places for neighbors to exchange information and discuss their local schools. Recycling and reuse centers operated by grassroots organizations deliberately expanded their community and environmental agenda and began calling themselves "community material reuse centers," "community recycling centers," and "community material resource parks." They have begun taking on social and ideological traits, developing a new culture and new rituals of recycling.

Wastewise in Georgetown, Ontario, and Arcata Community Recycling Center (ACRC) in California are only two examples out of dozens of community-based environmental organizations in North America that offer used material resources to their surrounding communities. Wastewise, the more recent of the two, opened in 1991 after local opposition to numerous large-scale waste disposal proposals. It is a registered nonprofit and a model for merging charity, sustainable economic development, and waste management with an environmentally responsive lifestyle. Its large warehouse accommodates recycling and reuse functions, as well as a wide range of educational programs—seminars, presentations, an environmental library, and a small display area. It relies especially on garage sales, resale of used materials, and donations. The membership program and fund-raising Christmas parties help maintain it as a viable business and a vital communal center. It is a place to socialize and to browse, tinker with, and recirculate reusable items. The center publishes recycling guides and a monthly newsletter, *Wastewise Watch,* that covers community events, publishes members' personal stories, and provides recycling tips.

The small university town of Arcata, California, containing 16,000 garbage producers, opened its facility in 1971, as part of the first wave of nonprofit community recycling centers. Arcata gained national attention for its pioneering, sustainable approach to solid waste, composting, and wastewater management (Riggle 1994, 78). Unlike Wastewise, the ACRC considers itself entrepreneurial, providing a value-added service that centers on waste management while promoting environmental awareness. With a crew of volunteers and donated trucks and labor, the center expanded its operation and slowly became self-sufficient. Packaging and containers are separated under a shed in the facility's yard, processed at the warehouse, and shipped to other sites to be used as raw material. Organics and yard waste are accepted for chipping and composting; a self-service oil change station is also available. Construction items and building materials are organized and displayed in a back yard for sale or exchange. But the highlight and pride of the site is the Reusable Depot, where books, clothing, packing, and household items are accepted for resale (fig. 4.5). The rummage items and used building materials attract many locals, who leisurely peruse the goods.

Arcata is only one of many community recycling and reuse programs along the environmentally conscious West Coast. In fact, whole "towns" have been created around the activity. Garbage Reincarnation Inc., a California nonprofit that has been providing various recycling services since 1971, started two Recycletowns, in Petaluma and Healdsburg. Recycletown Petaluma, the main installation, opened in 1986 and is located at the county's central landfill. Alongside its low-tech recycled pole barns, Recycletown had an electronic "bottle wall" until the town had to relocate when the county expanded its landfill. Now in another strategic location at the landfill's periphery, it continues to accommodate an enthusiastic public while awaiting a permanent home. The *Sometimes Monthly Recycle Rag,* Recycletown's official newsletter, contains an event calendar and various articles on the environment. The highlight of Recycletown is the annual Scrapture Competition and festival on the last Saturday of June. Everyone is invited to a celebration of creativity in which free junk food, great music, and theater accompany the creation of sculptures made of scrap. The event demonstrates the truth of Recycletown's motto: Garbage is the result of lack of vision.

As they combined waste consumption and redemption, many community-based reuse centers began to attract the same clientele that frequents the mall. Partly modeled after flea markets, thrift stores, theme parks, or shopping malls, reuse centers provide spaces for the familiar rituals of vendor and buyer in a commensal marketplace and take on a sense of social centrality in a less sterile and homogenized setting than the mall. People come to encounter unfamiliar textures, uncontrolled smells, and unexpected items (maybe even something they had once owned) in a convivial, relaxed atmosphere. Above all, various environmental programs, lectures, demonstrations, and exhibits set the collective ecological mission in motion.

Urban reuse social programmers

In large cities, entrepreneurial nonprofit organizations have tailored programs to inner-city needs. Known by names such as "surplus exchange," "resource reuse," and "material reuse," these operations make wise use of waste resource while creating jobs, targeting resourceless urban cultural and ethnic groups, bridging the gap between source and demand, and enhancing the quality of urban life. These institutions work to make their environmental agenda relevant to the many urban residents for whom environmental concerns are otherwise a luxury. One of the earliest of these organizations, Material for the Arts (MFA), linked waste diversion with cultural needs and created a treasure trove for its select clientele of New Yorkers. Since 1979 it has used its two trucks to pick up outdated high-tech equipment and used items from donors— major corporations, small businesses, and individuals. MFA houses the donations in a warehouse and offers them free to more than 1,300 registered art and cultural nonprofit groups—public schools, museums, musical groups, performance centers, and social, community, and health centers.[9] Other similar programs have since opened throughout the country.

In the 1990s many urban nonprofit reuse organizations started, generating hundreds of jobs and broad public interest. The Green Institute Reuse Center in Minneapolis is one example. Located in the impoverished Phillips neighborhood, the institute considers itself an entrepreneurial environmental organization and aims to pursue economic, environmental, and social gains simultaneously. In 1995 the organization, initially formed as a result of community opposition to a prospective garbage transfer station in the neighborhood, launched its first project, the ReUse Center, a retail store selling salvaged building and construction materials in an innocent-looking strip mall building at the Hi-Lake Shopping Center. Later, the Phillips Eco-enterprise Center was added; it offered commercial and industrial space equipped with state-of-the-art environmental technologies, including geo-exchange heating and cooling, active daylighting, and a green roof. The institute also manages the De-Construction Services, which reclaims salvaged material for future sales at the ReUse Center. Outreach programs teach home improvement skills and support neighborhood-based projects that focus on local green space, pollution prevention, and sustainable transportation. The green space program has already supported twenty community gardens and has planted dozens of street trees.

Other community programs further simplify and decentralize large reuse operations. Counties set aside areas for citizen drop-offs next to landfills or transfer stations and then sell the donated items. In the transfer station in Lane County, Oregon, a few covered booths next to the drop-off recycling containers offer a pleasant display of artifacts that people bring in on their weekly trip to get rid of recyclables. Charity organizations and stores add their trailers to such sites and collect a share of these goods. Some communities have instituted curbside reuse days. Once or twice a year, residents put out personal rejects that their neighbors look through for lucky finds, creating a citywide garage sale. In California alone more than a hundred such events have taken place every October since 1997, during Second Change Week, as it is called, putting back into circulation more than 100 tons of reused goods and material in October 1998 (Block and Wood 1998).

Programs that recycle organic waste add another dimension to the recycling landscape. They have not only created compost for sale but also transformed the urban landscape. In the early 1980s, cities began instituting compost demonstration yards to teach residents how to do their own composting. Then, public outreach programs, such as the Bronx Green-Up and Manhattan's Green Guerrillas in New York and Turn a Lot Around in Chicago, started reclaiming vacant lots in large cities, turning them into vegetable gardens and pocket parks for the community. The decentralized and locally tailored projects of community grassroots and private non-for-profit recycling organizations embody the spirit of supermodernism.

Artists and designers embraced the reuse of waste long before it was accepted and adopted by municipalities and the general public. Back in 1969, Patricia Johanson proposed to reclaim organic waste as well as urban wastelands in the from of productive, educational, and functional landscapes within the city. "Garbage Garden" (1969) uses organic waste as a proper medium for urban agriculture—as fertilizer for soil, biomass for sculpted earth, and mulch for retention of moisture. Johanson's proposal imagined roadsides, vacant lots, rooftops, and apartment terraces into animated urban gardens that feed off and thrive on city garbage.

Economics and creative imagination have been crucial in the process of reuse. The poor and artists, as contrasting as it may seem, have always found value in material discard, but their practice was hampered by conflicting institutional interests and aesthetics.

The Aesthetics of Recycling and Reuse

Filthy pests

Scavenging, which has been an essential element of the reuse system since industrial processes first gave refuse some value, was always considered a grimy trade, a threat to genteel sensibilities and good order. Once a thriving vocation, it was forced to the margins and largely eliminated in the early twentieth century. The artist Mierle Laderman Ukeles once said in an interview that she would like to make a movie that depicts a garbage truck collecting residential waste while escorted by an armored truck, like those that transfer money between banks—complete with armed guards (Yung 1996, 27). Perhaps Ukeles's fantasy will never be realized, but now that recyclables have value, the poor sometimes snag refundable cans and bottles from loads on the curbside awaiting early morning collection. Legislation, which has often been deployed to coerce and normalize the ethic of recycling and to sanitize and "aesthesize" the treatment of waste, is putting an end to this practice. Pressured by annoyed residents and savvy waste management companies, cities like Los Angeles have passed ordinances declaring curbside recyclables the property of the companies to enforce an orderly corporate collection. Today's trash can and curbside pickers recall the ragpickers of the past, and the same genteel mentality behind the ouster of today's scavengers was responsible in part for driving away those of the nineteenth century.

The nineteenth-century scavenger held a contradictory place in the European social imagination. The chiffonniers, as they were called, were viewed as deviants, collecting anything from stale bread and dog and cat shit and engaging in criminal activity. At the same time, they were considered creative beings, dedicated to making use of society's sweepings. Charles Baudelaire described the ragpicker as an archivist and a cataloger who makes intelligent choices while sorting through garbage. Parisian Bohemians were fascinated by the ragpickers, with whom they felt a partial affinity. Walter Benjamin (1973) made this point in his interpretation of Baudelaire's work: "But from the littérateur to the professional conspirator, everyone who belonged to the bohéme could recognize a bit of himself in the ragpicker. Each person was in a more or less obscure state of revolt against society and faced a more or less precarious future" (20). Benjamin also drew an analogy between the ragpicker's work and his own work as a cultural historian. The Parisian authorities were less impressed; led by Eugène Poubelle, they put a stop to ragpicking and instituted the garbage can (nicknamed the *poubelle*), which was to be put on the street only fifteen minutes before collection. Following the riot of the tens of thousands suppressed ragpickers, Poubelle allowed the trash to be traded in a confined area, which later became the famous Parisian flea market.

Describing the American scene, Susan Strasser recounts that until the early twentieth century, in major American cities hundreds of scavengers, commonly immigrants and the poor, women and the young, found a living in the trade of waste, a profession of borderline legitimacy (1999, 115, 117). Some were well organized, covering defined districts of the city. The ragpickers and bonepickers scavenged the streets selecting objects for which there was a market; rag dealers bought directly from households, and others picked through municipal dumps, at times paying the city a weekly fee for the privilege. Items were traded for cash, reaching the junk dealer directly or through a series of middlemen. Scavengers lived under bridges and in improvised shelters near their source of living. In many cities, claims Strasser, rundown streets bear such names as Bottle Alley or Ragpicker's Row after their former inhabitants (1999, 117).

In Hygeia, the imaginary city of health, Benjamin Ward Richardson envisioned a system of complete waste utilization: "The inspectors of the sanitary officer have under them a body of scavengers. These, each day, in the early morning, pass through the various districts allotted to them, and remove all refuse in closed vans. Every portion of manure from stables, streets, and yards is in this way removed daily, and transported to the city farms for utilisation" (1876, 41). New York's Colonel Waring, who put this vision into practice and institutionalized the recovery of household waste, was also the man who put an end to the effective diversion of useful items by New York's street scavengers. Scholars Jesse Walker and Pierre Desrochers see Waring's attitudes as "typical of the nascent Progressive movement: He confused neatness with cleanliness, and coercion with order" (1999, 74). (They also fault today's curbside recycling as being a moral imperative, not an economic one.) The institutionalization of the trade was signaled by the first publication of the

Waste Trade Journal in 1905 and by the founding of the National Association of Waste Material Dealers in 1913 (the organization later expanded and established various specialized divisions).

During the golden age of the junkman between 1910 and 1930, as society grew affluent and discarded more and "better" garbage, scavengers were gradually chased away. Competition from charitable organizations and the bigger businesses that came to dominate the salvage trade at the turn of the twentieth century dwindled scavengers' source of income. Moreover, municipalities took notice of the improved waste and sought to capture its wealth; they hired immigrants to pick through the collected garbage and sold the recovered material. More than anything else, however, the scavengers' stigma of being disorderly, "filthy pests" provided a good reason for their elimination. Influential municipal reformers and civic organizations, especially women's associations, started many initiatives to clean the city of its dirty matter and people. By the mid-twentieth century, scavenging had become extinct in most developed countries, left to sanitation workers (or to those lucky few who beat them to the garbage bins).[10]

In developing countries, however, salvage and scavenge waste trades still flourish. These have primarily relocated close to dumps where garbage trucks regularly unload. An exceptional model was put to work in the renowned green city of Curitiba, Brazil, where everything from transit, to industry, to garbage is conceived and managed with sustainability in mind. The city's 1.6 million inhabitants separate organic from inorganic waste and recycle almost half of their garbage in several plants. Yet the success of the recycling rate is credited especially to children and the poor, who are encouraged to collect and exchange material for food and basic necessities. The streets remain clean, the poor earn a living, and new recycling industries mushroom. Similar "integrated" waste programs that employ scavengers are being tested in other places.

Mongo, bricolage, and garbage housing

Whereas professional and opportunistic scavengers seek out monetary value from garbage, others search for castoff objects of aesthetic and symbolic value. Even sanitation workers refuse to allow certain things to pass and retrieve objects from the flux of garbage en route to the dump or recycling center. "Occasionally an object speaks to a sanitation worker in the same way it does for an artist," says Mierle Laderman Ukeles, who has spent much of her career with workers at New York's Department of Sanitation (interview with author, New York, March 16, 1999). Ukeles likes to speak of "mongo," a generic word they use to mean "value." In particular, mongo is the value assessed in an instant as an object is on its way to the truck. It is conferred on certain religious objects, such as statues and religious icons, and on artworks, such as painting and sculptures. Many of the items are saved and displayed on walls and shelves in workers' lounges or in storage spaces of the Department of Sanitation. At some recycling centers, workers picking through the fresh garbage on conveyor belts extract numerous treasures and mementos, including wedding rings and currency (sometimes hundred-dollar bills). In workers' offices at landfills, one is also likely to find a collection of certain objects retrieved from the dump. The manager of the Hiriya dump in Tel Aviv proudly exhibited his coin and gun collection to visitors, and the workers of Haruvit dump, near S'dot Micha filled their trailer office floor to ceiling with sexual paraphernalia, including pornographic magazines and images, and children's toys, such as dolls and teddy bears, all found in the dump.

Folk artists' creativity, imagination, and manual dexterity make them true scavengers. They have always found some value and beauty in waste and put it to service, creating one-of-a-kind objects. The French called this practice *bricolage,* meaning tinkering or the work of a putterer. The *bricoleur* assembles tools, materials, and objects for future use and considers projects related to and based on what is at hand (Strasser 1999, 11). The resulting new constructions arise from an intimate knowledge of the available material. Much like the bricoleur, though free of utilitarian purpose, folk artists have used salvage material to construct follies, fantastic structures of idiosyncratic and enigmatic meanings in their private yards. Howard Finster's Paradise Garden in Pennville, Georgia; the Rock Garden of Nek Chand Saini in Chandigarh, India; and, of course, the famous towers built by Simon Rodia in the Watts neighborhood of Los Angeles are but three of many well-publicized examples of such constructions. When Rodia finished the tantalizing cluster of towers, he gave the land and all the art that was on it to the city. Following a long fight to save the towers from the city's intent to tear them down, the Committee for Simon Rodia's Towers in Watts (a group of concerned citizens that recognized the value of the work) won the protection and designation of the site as a National Historic Landmark and a State Historical Monument. It has since become a major tourist attraction and a center for art and jazz festivals.

These prodigious builders undertake bricolage as a hobby. It provides them with an opportunity to establish identities as creative artists. Venri Greenfield, who interviewed folk artists who make things out of junk in their yards and homes about their motivation, found that recycling is spurred by some economic, utilitarian, artistic, or political motive, alone or in combination with others. Some of the artists spoke of achieving an independent life, others of countering capitalistic consumerism or avoiding standardization and specialization; still others stressed that creative, aesthetic activity aided their personal growth, self-expression, or anxiety reduction (1986, 2). In the hands of vocational reusers, discarded objects still retain the memory of past use and become vessels for new meanings. Folk artists find new forms in the random mix of salvage and suggest fresh, personal interpretations of the original object. They are able to disassociate an object from garbage—its smell, ugliness, pollution, chaos, and death—and reassociate its formal, material, functional, or symbolic attributes with a potential use. Folk art has more recently infiltrated the high art institutions and permeated the discourse of art, a change that reveals much about the art world's own beliefs and practices.

Like bricoleurs, architects have experimented with structures of salvaged waste to solve social and environmental problems. Martin Pawley and Mike Reynolds have made a systematic effort to employ disused materials in affordable and environmentally sensitive buildings. Since the early 1970s they have experimented with house prototypes from discarded tires, crates, beverage cans, and bottles. Reynolds took a hands-on approach and experimented with new techniques, such as using dirt-filled old tires as a foundation for his famous Earth Ship home (Wilson 1993, 4). Pawley has been a harsh critic of current industries that produce and recycle material. He wrote a series of well-received books, including *Garbage Housing* (1975) and *Building for Tomorrow: Putting Waste to Work* (1982), which explore the possibilities of engaging the massive stream of waste materials in cost-effective and sustainable building construction. Pawley stresses the need to give secondary-use structures genuine and dignified identity in order to avoid the stigma of secondhand material; such identity, he believed, is crucial to their acceptance by the mass-consumer culture.

Both Pawley and Reynolds participated in the first and only International Conference of Garbage Architects in Florida in May 1979, where the issues of low-cost dwellings and conservation practices related to housing were raised (Pawley 1982; Wilson 1993, 2). In the 1980s, however, housing for the poor was forgotten, with or without bricolage. Bricolage remained merely a source of romantic allure and inspiration, but it had no impact on popular architecture. Only a few architects ventured into the secondary use of building materials—primarily in low-cost, experimental housing, and often in nonconformist high style.

Recycled art

Artists (with a capital A) have long drawn on the cultural significance and aesthetics of waste and reuse. They have a natural tendency to engage intellectually and aesthetically with the detritus of human activity. As agents and transformers of cultural consciousness, they rethink the material and cultural essence of castoffs and propose ways to reimagine them. Although some artists reuse discarded items in their art out of necessity, most find in such objects an array of stimulating possibilities. Some artists use discards simply as pigments for painting, for their visual or textural attributes. Others dwell on their nostalgic and associative traits. Others employ them to declare their avant-garde tendencies and provide social commentary. As early as the 1910s, collage work in painting began taking on the characteristics of bricolage, as Pablo Picasso and Georges Braque began combining images and newspaper fragments into their work. These gestures brought everyday material reality into the artistic context. Following the cubists, dadaists and surrealists such as Ernst, Magritte, Picasso, and Braque took to its logical conclusion the idea of collage as an assemblage of displaced, unrelated contextual images. Students at the Bauhaus design and art school were encouraged by Johannes Itten "to keep [their] eyes open, while out walking, for rubbish heaps, refuse dumps, garbage buckets, and scrap deposits as sources of material by means of which to make images (sculptures) which would bring out unequivocally the essential and antagonistic properties of individual materials" (Rosenberg 1989, 61).

Found objects also entered the central production of high art in the 1910s, as their use in sculpture and installations was popularized. Marcel Duchamp legitimized the reappropriation of "readymades" (ubiquitous, mass-produced objects) with his *Bottle Rack* (1914) and *Bicycle*

Wheel (1913). Pop artists continued to address the proliferation of short-lived, mass-produced objects and the aesthetics and politics of consumer culture in the 1950s. Arman piled up used telephones, discarded cigarette butts, and used cars in his installations. Ed Kienholz articulated political messages in large-scale installations, continually interrogating the possibilities of reformulating new meanings out of used, displaced materials.

When displayed in a public space, works made of junk elicit varying reactions and opinions, ranging from delight to contempt. The transformation and contextual shifts that artists-recyclers achieve may be perceived differently by viewers whose familiarity with the raw materials influences expectations. The artist Nancy Rubins makes large public sculptures of scavenged objects, such as domestic appliances and even mobile homes. She uses discards she finds wherever the piece is to be installed or those referencing elements at that site in order to draw attention to local waste. Her gigantic sculptures seem to erupt, as if they are about to consume their surroundings. Her attempts to confront people with their own waste have often elicited passionate opposition; some of her public installation have been mistreated and scorned. Rubins's first junk-made public sculptures enraged the public and were rejected. In Chicago, *Big Bil-Bored* (1980), a massive wall of toasters and other discarded domestic goods at the Cermack Plaza shopping center, was voted in a radio poll the ugliest sculpture in the city and condemned by a mayoral candidate as dangerous to the public health. In Washington, D.C., Rubins created *Worlds Apart* (1982), a 45-foot-tall assemblage of discarded appliances alongside the Whitehurst Freeway (fig. 4.6). It was ridiculed on television programs and created a public outcry about wasted tax dollars (Duncan 1995). These responses reflect the threat that garbage still poses to us, as well as the fine line dividing junk from scrap art. Sometimes the difference is in the details and craftsmanship; at other times, it is in the context.

Contemporary artists have increasingly immersed themselves in and explored society's residue as a way to approach issues of identity and social norms. "Garbage is the new anti-aesthetic," claims art historian Jo Anna Isaak, listing eight major exhibits about trash, dirt, and shit that opened in the 1990s (2001, 174). These artists' work continues to expand the realm of aesthetics (the aesthetics of leftovers and rejects) but remains highly controversial. The most provocative junk artist today,

Tomoko Takahashi, has garnered both high acclaim and sharp criticism for her installations. She often collects the daily refuse left in an office or waste from a demolished structure and rearranges it into a meticulous clutter without any coherent order (Buck 1998, 121).

Recycled environments

Beginning in the 1970s, another group of artists engaged trash and recycling environments specifically to raise environmental consciousness. Mierle Laderman Ukeles, who perhaps deals most bluntly with issues of waste and recycling, has labored for many years to expose the mindlessness with which society treats its resources and to rethink our relationships with materials, maintenance, and the sanitation workers who serve us. Ukeles used discards from the New York Department of Sanitation—outdated machinery, workers' equipment, and materials from the waste flow—in many installations at the city sanitation facilities. In *Touch Sanitation Show* (1984) at an old marine garbage transfer station, she displayed a sequence of heavy-duty dump trucks and machinery, workers' equipment, and metal mesh bins, each containing recyclable waste. In *Re-entry* (1987) at P.S.1 Contemporary Art Center she created a 45-foot-long passage lined with recyclables—steel gratings, crushed glass, and aluminum cans—accompanied with taped soundtrack of sanitation trucks in action.

"Lobotomy" is a term Ukeles uses to describe the exploitation of resources, first during their extraction and design and then in their disposal. She declares: "We are using Earth's resources without designing their release into an ongoing flow of circulation. When we have blockage in our bodily flow we die" (interview with author, Tel Aviv, January 3, 1999). Her interest in recycling resides in the notions of flow and time, the trajectory of each material through its life span. She urges people to imagine waste facilities as the great public art of our age (1995, 193). In her most intriguing project, *Flow City* (1983–96), which was never fully realized (see below), Ukeles reshaped the 59th Street marine garbage transfer facility in New York into a public working museum.

The idea of flow and flux is now shaping the architecture of some garbage facilities, such as garbage transfer stations that function specifically as terminals of flow. In Zenderen, the Netherlands, for example, the architects Oosterhuis Associates built a gigantic, gracefully curved structure in 1995 (fig. 1.3). Its 469-foot-long ellipsoid volume houses all its functions under one roof—offices and computers at the front, waste handling and sorting at the center, and leachate and gas purification from the adjacent landfill at the rear. The column-free space is designed to be converted into a sports or cultural facility when it is no longer needed to transfer garbage (Pearson 1998).

In designing new transfer and recycling buildings, artists and architects also make critical decisions on plant sitting, public programs and access, and the connection to the community. The most prominent precedent is the 1993 27th Avenue Waste Recycling and Transfer Facility in Phoenix, Arizona, designed by Michael Singer and Linnea Glatt (see below). The Seattle environmental artist Buster Simpson proposed to use the percent-for-art fund—money from a public program that dedicates 1 percent of a project's total cost to public artwork on its site— to open a secondhand goods store and a gallery / workshop in one of Seattle's planned waste transfer stations. In *Poetic Utility*, his artist-in-residence proposal for the Seattle public utilities, he explains: "The gallery is a mix of pedestals for unique 'ready mades' or recycled works of art. These pedestals could be sited along the public edge of the facility or flank the drive-thru route. Agit props in the form of work on billboards could convey issues consistent with the facilities mission" (1998, 12). Yet other projects call on artists simply to embellish existing facilities, to make them more palatable and decorative. In the recently completed Vashon Transfer / Recycling Station in King County, Washington, Deborah Mersky used percent-for-art money to create a porcelain enamel panel for the weigh house booth and an aluminum frieze for the recycling drop-off wall. Tom Brennan and Jim Pridgeon redesigned the fence of the Enumclaw Waste Transfer Station in southern King County and located symbolic sculptures inside the facility grounds.

Artists have also been effective organizers of recycling programs and events. Many parades with floats and decorations made of recycled materials have kept artists busy. Buster Simpson has proposed to create a "counterdisplay" at the popular Seattle Home Show, where artists would exhibit alternative products. Many art exhibitions intended to help broaden concerns about waste issues and normalize the act of recycling came on the scene during the 1990s; they were housed in environmental centers, inside waste facilities, or in cultural centers and science museums. Two examples are the 1993 *River of Resources* exhibit in the Resource Recovery Authority Visitor Center in Hartford, Connecticut, and the 1998 *Rotten Truth (About Garbage)* exhibit in the Montshire Museum of Science in Norwich, Vermont. These are typically didactic and entertaining displays with factual information and interpretation about garbage hazards, landfills, recycling, and recycled products.

In contrast, a 1996 traveling exhibition named *The Museum of Garbage,* designed by Monika Gora and Gunilla Bandolin, was a highly stimulating and reflective project that was housed in a truck trailer and circulated for a number of years through various Swedish towns. Down the middle of the trailer space six glowing cases containing thirty objects extracted from a landfill moved along a conveyor belt. Six computers showed animated stories, composed by various artists, about the objects' life, identity, and connections to other phenomena and materials. Along the sides of the trailer, several displays of garbage mess invited viewers to touch them using rubber gloves. At the far end, a film ran about the process of the extraction. The exhibition was intended to awaken the imagination about daily objects and materials in the same way that a picture or sculpture does, to draw attention to "the dreams and hopes represented by ordinary objects, and what happened to those dreams when the objects pass out of our lives" (Gora and Bandolin 1996, 70).

The ingenuity with which artists, architects, and folk artists relate change, flow, and the redemption of waste can serve as an inspiration and a model for social-environmental change. Not many artists offer practical solutions, but many open up dark and hidden spaces and challenge and reform our aesthetic sensibilities. They seek justice and transparency in the system that makes key decisions about waste production and management and facility siting. Although they may not concur with mainstream desires or dominant aesthetics, they fight for quality architecture and respect for local communities. Designers and artists rethink (rather than reapply) recycling, reconsider people's relations to the material dimension of things and places, and, when allowed, reexamine recycling and reuse facilities in terms of how they express and enhance culture.

Recycled Misopportunities and Opportunities: Projects and Critique

Since the 1980s several highly publicized and celebrated solid waste reuse, recycling, and transfer centers have been built. They offer valuable lessons, highlighting the opportunities such facilities provide, as well as their limitations. Beyond their reuse and recycling functions, these facilities make an effort to provide public access and education, yet they greatly diverge in their political, social, and environmental focus.

Projects concerning the design of recycling facilities fall into three groups—privately owned, publicly owned, and grassroots and / or nonprofit. Sanitary Fill in San Francisco and the Recyclery in Milpitas, California, which house a sculpture garden and a recycling museum respectively, represent the first group. These private corporations use small, physical, seemingly educational gestures to appease adjacent communities, satisfy popular "green" public sentiments, and project the image of having an environmental raison d'être. Municipal projects are broader in scope and purpose, commonly promoting the educative and celebrative approaches. *Flow City,* Mierle Laderman Ukeles's garbage marine transfer station in New York, and the 27th Avenue Solid Waste Recycling and Transfer Facility in Phoenix, Arizona, both designed as publicly accessible urban landmarks, represent the second group. Public projects offer artists and designers greater freedom and enable them to create civic architecture and shape how waste facilities are perceived.

Rather than imposing restrictions, as do government facilities and for-profit corporations, grassroots-initiated and nonprofit recycling and reuse organizations open up an unlimited range of possibilities for addressing social and environmental problems, as well as creating productive and vital public spaces for the community. Three facilities, all of which combine sustainable and integrative approaches and embody the practical and social ideals of supermodernism, represent this group. The recycling facility in Wellesley, Massachusetts, has forged a contemporary version of a town commons. The Eugene, Oregon, facility called BRING has created an environmental motivation center in the form of a junk-yard adventureland or reuse theme park. The Resource Center of Chicago, a multifaceted social and environmental program, has tailored a job and service center for inner-city poor and aims at economic and urban recovery. While art and design are scarcely used to shape the last group of places, which take on a kind of anti-aesthetic tone (or pluralistic aesthetic), they are intended to stimulate creativity and imagination, key elements in adaptive reuse.

Private gestures

Private initiatives are limited in scope and purpose. As they engage community concerns they also advance conflicting interests (i.e., profit), of-

ten employing artists and designers to serve corporate agendas, to facilitate public relations, to function as mediators between the community
and the business, and to make facilities aesthetically palatable. Within
these confines, the role of art and design cannot challenge and critically
address the issues at stake.

Sanitary Fill Solid Waste Transfer and Recycling Center. Located in an industrial fringe on top of a closed landfill in southeast San Francisco, the
Sanitary Fill Solid Waste Transfer and Recycling Center (a subsidiary of
Norcal Waste Systems, Inc.) started its operation as a conventional transfer station in the late 1960s. The plant today reclaims more than a dozen
different materials from sorted and unsorted waste, including toxic
waste. Although this long-established facility is far from an impressive
or innovative model for recycling, it has acknowledged its obligation to
the nearby community (and its own employees) and made a significant
gesture of goodwill.

The rather plain-looking garbage recycling and transfer facility, with
spare utilitarian, horizontal sheds frequented by trucks and private vehicles unloading and loading waste material and recyclables, is host to a
special garden and an art program. The three-acre sculpture garden
named River of Hopes and Dreams occupies an edge zone abutting the
lower-class neighborhood Little Hollywood and is known only to workers, neighborhood residents, and visitors who are shown it on organized
tours. Jackie Tripp, an artist hired to beautify the plant, realized the opportunity for community outreach and proposed a garden and an artist-
in-residence program. The garden was designed and built in 1992 by the
artist Susan Steinman together with a group of local high school students
(Baggett 1992, 34).

A small sign at the garden entrance reads, "The sculpture garden features works from artist-in-residence program which challenge old habits
and raise awareness of environmental priorities concerning waste mini-
mization, re-use and recycling." A sinuous, light-blue-colored concrete
path carved with people's names and wishes leads the visitor from the
entrance through flower beds of bursting, purple African lily to an
arrangement of rocks from which the path seems to emanate like a river
(fig. 4.7). The San Bruno Mountain range serves as a backdrop. Large concrete pavers dispersed in the central lawn on both sides of the main path
are embedded with various fragments and materials found in the garbage. Sculptures made by artists in residence from retrieved junk bottles,

scrap metal, and other familiar objects stand amid the garden's flower-beds and paths.

This island of respite from the plant's frenzied operation serves as a place for employees' lunchtime and breaks, for organized picnics, and for educational programs and tours. It also provides a buffer for residents in nearby homes, alleviating the sensory nuisances and reducing mental strain. Unfortunately, access to the garden is restricted, and there is no direct entry from adjacent streets. As it is now, the garden encoded with associations of life, health, and morality draws attention away from the working plant, conceals it from neighboring houses, and wipes out guilt over using technology. The plant and the garden stand as opposites of work and leisure, ugliness and beauty. Seemingly balancing one another, they remain separate and alienated entities, deprived of any reciprocal dialogue.

The company would undoubtedly make a more meaningful contribution if it opened the garden to the neighborhood and established a community art program, for example. Local schoolchildren and residents could have made creative use of waste materials recovered from the facility and might have enlivened the park with new additions.

The Recyclery. While Sanitary Fill employs a garden and an art program to project an image of social and environmental responsibility, the public relations tool of the Recyclery at Milpitas, California, is a museum and information center. Built as a gateway to the Newby Island Landfill, the Recyclery is a filtering station; there, all recyclable materials are removed from the waste stream so they do not end up in the landfill. Its practical justification is to save landfill space, thereby extending the landfill's operating life.

Owned by the Houston-based Browning-Ferris Industries, Inc. (now part of Allied Waste), one of the largest multinational waste disposal companies in the late 1980s, the company has ridden recycling's wave of popularity in California. The exhibit at the education center features interactive electronic displays that focus on the institutional view of problems caused by solid waste and on the role that businesses and private citizens can play in solving them. A dramatic 100-foot-long "wall of garbage" represents three minutes' worth of Santa Clara County's refuse. The display seems to target schoolchildren and delivers simple, factual information through buttons, levers, and posters. A buyback center is housed in the same building, and individuals who arrive to redeem recyclables can take in a short visit at the center as well. An outdoor path marked with blue grizzly bear tracks leads from the exhibit hall to the main building and into a small, raised viewing deck that looks out over the operation inside the Recyclery's main structure. Indeed, the observatory is the most revealing point of the facility, a window into the real working museum. The plant's relationships to the sanitary landfill and the bordering wetland and bay are unfortunately kept inaccessible and thus invisible.[11]

In an effort to improve their public image and link themselves to the green ethos, privately owned waste facilities have opened their doors to public tours and to art and educational programs; unfortunately, they generally do so to serve their own needs and purposes. This has prompted some avid environmentalists to condemn recycling as merely another kind of consumption and recycling propaganda as corporate greenwashing that advances an institutional agenda rather than the common good. Art and design in private projects have yet to challenge the core assumptions of their patrons, the corporations that monopolize the waste and recycling scene.

New civic monuments

Under the leadership of enlightened city governments, art organizations, and visionary artists, a few public waste recycling and transfer projects have opted for a more significant mediating role between people and waste infrastructure, producing grand civic monuments. They are shaped as memorable, informative, and exciting places to visit and work, and they are integrated into the context of the city. Sometimes, however, the politics, legal responsibility, and inadequate budgets of large government projects doom these facilities to failure.

59th Street Marine Transfer Station/Flow City. The redesign by the architect Richard Dattner (with Greeley and Hansen Engineers) for the 59th Street Marine Transfer Station of the New York City Department of Sanitation, a place where city garbage is loaded onto river barges (whose destination, for many years, was Fresh Kills landfill on Staten Island), is simple but elegant. The structure is shaped to match the industrial language of other Hudson River waterfront piers, with an elongated boxlike shed for the barge lane and tipping floor and a covered bridge, inspired by regional examples, for the ramp leading to the tipping floor. A gate replicating the facade of a nineteenth-century sanitation pier stands at the entrance and bears the name of the facility.[12]

The involvement of the New York Arts Commission in this project ensured not only quality architecture but also an artist's input. In 1983, when design work began on the new transfer station, Mierle Laderman Ukeles proposed that a public art environment be incorporated into the building's structure. For the first time in the city's history, new zoning regulations were passed to allow public access to what had formerly been off-limits as a municipal workplace. Even though the adjustments were made to allow public access, the facility has never been opened to the public. Ukeles's site-specific project *Flow City* (1983–96) represents a museum very different than that of the Milpitas Recyclery; rather than offering solutions and public relations, it emphasizes questions and public substance. It avoids interactive games, simulations, didactic signs and information, and any techniques of beautification. Instead, it boldly and genuinely exposes the facility's ongoing operation. "The geometry and the colors are just spectacular. It's the violent theater of dumping," says Ukeles about the station (Doran 1996, 213).

Ukeles' design transforms the facility into a stage and the operation into a performance, and it engages visitors as they travel along a carefully planned path through a series of sequential, participatory environments and observation points. Though the experience was to be mediated and people were to be guided and oriented and their movements choreographed, no screening or buffer zones were included. Ukeles designed a long entry passage, running parallel to the ramp used by the garbage trucks and surfaced with a spiral of twelve different recyclable materials into which forty taillights from defunct garbage trucks were embedded. Those walking through the passage would hear recorded noises of the facility in operation, and they would reach a bridge enclosed with glass from which the actual dumping of waste from trucks onto the tipping floor could be viewed (fig. 4.8). The city's skyline can be seen on one side of the bridge and the Hudson River on the other. A 10-by-10-foot "media flow wall" of thirty-five video and computer monitors was to be located at the other end of the bridge. It was intended to transmit the frenzied activities of dumping and moving waste from multiple cameras strategically located at the transfer station, the river, and Fresh Kills landfill (Gablik, 1989; Phillips, 1989).

Flow City enables visitors to see unaided the ongoing work of processing waste, and unlike displays at other waste facilities, it does not frame the experience with educational and environmental rhetoric. Ukeles simply reclaims respect for the waste place, transforming the otherwise maligned, noisy, smelly, disorderly facility into a site worth visiting. If it had opened, *Flow City* might have become a hot tourist attraction. Officials blame the halt primarily on a lack of funding for personnel needed to maintain public access and provide information, but according to Ukeles: "The sanitation department is not ready to deal with the public. They still are and feel as the other side, the dark side that should remain invisible" (interview with author, New York, March 16, 1999).

27th Avenue Solid Waste Management Facility. In contrast, the 27th Street Solid Waste Management Facility did allow public access, and it has become a major draw even for tourists. The public announcement of the 1993 opening of the Phoenix recycling and transfer facility attracted a few hundred curious and proud residents, who attended a picnic and dance at the newly built civic palace. A catalyst for the popularity of this plant was the Phoenix Arts Commission, whose mission includes integrating public art with infrastructure. A design team led by Linnea Glatt and Michael Singer worked together with the engineering firm Black and Veatch, which had initially planned nothing more than a respectable-looking utilitarian shed. Glatt and Singer envisioned a facility that gives priority to the education and experience of visitors. They crafted fresh solutions to site layout, employee space, public access, and construction details and developed a comprehensive program that encompasses a garbage transfer facility, recycling activities, self-haul material recovery, vegetation mulching, and education and demonstration programs.

The building is sited to align with a visual axis that connects with and directs views to the city. Cadmium-colored bougainvillea is planted along walls of gray cement block that rise in shallow tiers (fig. 4.9). The visitor is led across a steel bridge connecting the parking lot to a shady, lush courtyard abutting the main building and offices. The core building, a rectangular space the size of two football fields, contains large sorting and recycling machines that are installed with the precision of classical statues. The space is visually accessible from the main courtyard, amphitheater, and catwalks through large windows carved in the building's walls. The catwalk takes visitors to an observation deck inside the noisy and dusty workspace, where workers, garbage, and machines move rhythmically in unison. The architectural critic Herbert Muschamp describes the interior: "Translucent panels and skylights bathe the interior with Gothic light. The effect is not pretty. Instead, the artists have reached for awe" (1993, B31). The project has won multiple awards, and dozens of newspapers and magazines have praised it as a new model for the integration of architecture, infrastructure, and public art.

As part of a grand master plan that builds on the facility's program, the Center for Environmental Learning and Enterprise is currently being considered for the 26-hectare area surrounding the 27th Avenue facility (see chapter 3). When and if completed, the complex will become a nexus of activity centered on the wise use of resources and energy.

But the potential success of such large government projects can be undercut by political bickering, competing interests of waste management businesses, operating inefficiencies, and volatile conditions of recycling markets and the waste stream. Such was the fate of the San Diego Materials Recovery Facility at the foot of the San Marcos Landfill, which enjoyed nationwide publicity in the early 1990s and created high expectations. As designed by the architectural firm James Stewart Polshek and Partners, the waste treatment facility called for a multifunction operation that would invite people to recycle their household waste and at the same time visit a gallery and a visitor center and stroll through an open-air colonnade and monumental water garden supplied with on-site treated wastewater (Griffin 1987).

Another ambitious and fragile public initiative, which garnered attention and publicity throughout the 1990s, has also stalled. A new model of a paper recycling plant was slated for the 26-acre derelict Harlem River Rail Yard in New York's South Bronx. The Bronx Community Paper Company was intended not only to process a quarter of the paper generated by the city, save virgin forest, and reclaim a contaminated site but also to serve as a catalyst for the economic and social revitalization of its depressed neighborhood.[13] The artist Maya Lin worked together with Harris Group engineers and HLW International architects to design an elegant workplace responsive to both the community and the employees. The result, "visually spare but symbolically rich," emphasizes significant aspects of the recycling process (Muschamp 1998). Transparency and multiple vantage points for viewing the operation were key elements of the design.

In all these capital-intensive public projects, art and design encourage community outreach and help rescue infrastructure from being isolated and merely utilitarian. Issues of access, siting, pluralistic aesthetics, and richer programs need to be explored further. While those running both private and public facilities conceive of the recycling of waste as an industrial operation to be kept clean, tidy, and under control, to be viewed as spectacle if seen at all, small-town community recycling centers and nonprofit organizations use recycling and reuse as tools both for changing the environment and society and for infusing places with vital and productive social and economic activities.

Sites of transactions

Initiated and often managed by environmentally and socially concerned citizens in small communities or inner cities, reuse and recycling centers have extended their emphases—from conservation to broader environmental education to social action and enterprise. In these "resource material parks" and "sites of (trans-)actions," social and economic organization and political skills play greater roles than high design and art. The human dimension—a particular individual or group—is always an integral part of the success story of the place.

Wellesley Recycling and Disposal Facility. The highly publicized and celebrated disposal facility in Wellesley, Massachusetts, a wealthy and educated town of 27,000 garbage producers, is a place that "looks like a park! . . . a place residents take pride in . . . a place to meet neighbors" (as a town brochure boasts). The dump, as it is still called by locals, is an 80-acre shrine of recycling that has itself become a commodity. When real estate agents try to lure potential home buyers to the community, they drive them through the dump. Neatly organized and clean, it is a showplace for tourists, a stage for politicians, and a social and literary center for the community. People go there to get rid off things but stay to swap and socialize, picking over each other's trash and keeping up with the events of the day (Clendinen 1985).

Officially known as the Wellesley Recycling and Disposal Facility, it started its source separation recycling program in 1971 when the town incinerator at the dump was shut down to comply with new federal emission standards. A local environmental group called Action for Ecology (and later Friends for Recycling) began lobbying for trash recycling; it placed 55-gallon barrels at the municipal incinerator site, diverting some recyclables from the waste stream headed to another town's landfill. In time, the barrels were replaced with dumpsters, and the Department of Public Works took over the staffing and bought the old incinerator building that became a collection and transfer depot.[14]

Pat Berdan, the thrift-minded city director of public works who was brought up on the ethics of World War II savings and scrap campaigns, worked together with the Friends for Recycling to promote recycling education and ensure the facility's success. Berdan adopted recycling as a core strategy for the town and made it economical. He also envisioned the site as "a park, a place where people can become involved in environmental issues" (phone interview with the author, December 15, 2001). The new recycling center opened in 1995 (designed by SEA Engineers of Cambridge, Massachusetts), designed with an access road that loops around the old furnace building and leads to twenty-one spotlessly clean sorting bins and bays and five drive-through trash compactors. Colorful and informative signs direct residents to the proper bins for all types of glass, newsprint, papers, metals, and plastics. There is a Goodwill container for old clothes and areas for household items, tires, appliances, building material, and used oil. Leaves and grass clippings are composted and made available free to town residents, as is firewood.

Because Wellesley has never had a municipal garbage collection system, on Saturday mornings large numbers of townspeople congregate at the dump, a ritual in their busy weekend. Yuppie patrons drive through in their Volvos to drop off their empty Canada Dry bottles, go on to the Take It or Leave It section to inspect discarded television sets, and then browse through the Book Swap. Local and even national politicians seeking votes stop here to campaign; citizens, to voice protest. The shaded bookshelves, where English poetry and prose mingle with volumes on economics and religion, are clearly the high point of the dump (fig. 4.10). People come to deposit what they have read, browse, and take home what others have left. When they seek specific books, they leave request notes on the bulletin board for the librarian, a volunteer from the League of Women Voters, who will contact them when the book is found. The outdoor library area, a brick patio with planters and four benches made of recycled plastic, is designed as a convenient place to sit and read.

The site has evolved into a principal social center, complemented by a recreational park setting. There are picnic tables under trees, and people arrive in the spring to gather mushrooms. Not surprisingly, the site has been dubbed a successor to the New England town commons or market square. While the commons once stood for community as the place where individual herds grazed together, the recycling center does the same as the place where castoffs are shared. The *Chicago Tribune* refers to Wellesley's recycling center as "a logical manifestation of Yankee ingenuity and frugality" (Coakley 1986, A20). A *Wall Street Journal* reporter considers the dump as sensible and sophisticated as New England itself. He quotes a professor to provide psychological insight: "Going to the dump is 'a purifying ritual of the upper class. They seem to enjoy it because it's good, honest work.' . . . 'The dump serves the function of putting very rich people back in touch with the work ethic'" (Morrell 1998, NE3, 1).

BRING Recycling. Wellesley's clean and upscale center contrasts the junky and festive atmosphere of BRING Recycling in Eugene, Oregon. A grassroots, nonprofit organization, Begin Recycling in Neighborhood Groups (BRING) celebrated its thirtieth birthday in December 2001. BRING promotes reuse as a basic strategy for achieving a sustainable future and prides itself for luring to its junkyard style facility many area residents that are not "typical" environmentalists. Now that it has grown out of its small site and is about to move to a new location, its plans to become a major regional attraction—a kind theme park of reuse—are quite ambitious.

The center was started in 1971 as a pioneer bottle refund program by environmentally minded individuals who came to Eugene to found a commune. The group leased from Lane County 1.5 acres of swampy land at an industrial margin of the city, near the proposed Seavey Loop Sanitary Landfill site. It was the first recycling institution in the county. BRING began with a warehouse and invited people to drop off aluminum cans, tin, glass jars, refillable bottles, and steel in sorted barrels they provided. It then sold the items to the public and to manufacturers. Soon, a truck—the BRINGmobile—began collecting items from around town and a network of twenty-four self-service recycling depots sprang up.

BRING expanded its recycling list and in 1976 began processing the materials from all of Lane County at a central location, called The Dump. It also offered recycling information to citizens (taking up to 150 phone calls per week) and educational programs in schools. When, in the 1980s, recycling was becoming a way of life and municipalities began providing citizens with curbside recycling, BRING turned its focus to reusables. Today its twenty-one well-paid employees process and sell recyclables and reusables, provide a building tear-down service and workshops, and manage one of the most popular and extensive reuse warehouse on the West Coast. In 2001 at least 650 bathtubs and showers, 1,600 doors, 375 sinks, 500 toilets, and 1,000 windows were salvaged and put back to use. The organization is self-sustaining and projects that its operation will double by 2004 (BRING 2001).

BRING considers itself the city's recycling conscience. Bob Keefer, a local journalist, calls it "a junkyard with a mission, a sprawling, counterculture garage sale . . . selling junk and preaching the gospel of recycling" (Keefer 2001, H1). The facility, decorated with brightly colored signs and colorful flags akin to those on car lots, feels like a fair. As customers enter the gate, an employee at a stand greets them and offers advice. Booths, sheds, and a tent form a maze, and all are stacked, ground to roof, with meticulously ordered windows, doors, decorated hubcaps, toilets, stylish sinks, claw-foot bathtubs, lawn mowers, bicycles, and much more (fig. 4.11). Narrow gravel walks weave in between the items. Hundreds of artists, artisans, mechanics, builders, bargain hunters, hobbyists, and low-income residents come here seeking the perfect find.

Julie Daniel, the center's general manager and one of the founders, is as excited as ever about the future of BRING (interview with author, Eugene, Oregon, April 28, 2001). She envisions the new flat and dry 3-acre reuse park at a nearby Glenwood location as an active reuse and education center, filled with surprises and fun. Stacks of salvaged material will be housed in better and nicer structures along two sides of a main path, like the stores in a shopping mall, in order to become more user friendly and cater to middle-class sensibilities. Daniel imagines "Reuse Central" (a possible future name for the center) featured in the 2005 issue of Best Northwest Places in *Travel Magazine,* and highly recommended as a detour off Interstate 5: "From the moment you step through the stunning archway that marks the entrance to the site, you'll be delighted, surprised, intrigued, amused, entertained and yes, educated. BRING has managed a minor miracle, making the apt to be dull and preachy subject of resource conservation and recycling downright fun" (Daniel 2001, 2). Everything at the site, from paths to signs to structures, from the "recycled garden" that features creative reuse through demonstrations and installations to the "sound garden" filled with "wastruments," is made of salvaged objects. "How-to" demonstrations are held each weekend at the "rethinking center," where one can find a library and displays on recycling and reuse. Daniel enthuses, "On your way out, don't miss the Salon de Refuse, a boutique stocked with creations made from discarded materials by local craftpeople" (2). Other imaginative ideas Daniel has for the new site include a space for second (recycled) marriages and a box for recycled (environmental) confessions.

Damien Czech, the operations manager, is behind many of the creative assemblages that are already displayed at the site, as well as the

Reuse Recipes that are published in BRINGS's quarterly newsletter, *Reuse News*. The desire to cater to a large group of customers has shaped a model that is inclusive, egalitarian, familiar, light, humorous, and entertaining, blending reuse center with the most popular mass-culture venues, the shopping mall and theme park.

Resource Center. Rather than the popular and comic tenor of BRING, Chicago's Resource Center promotes a serious, social-environmental focus. It uses reuse and recycling as tools to recover material and human resources and to revitalize inner-city neighborhoods. It merges social and environmental causes into one. The center's five facilities are dispersed through the depressed neighborhoods of South Chicago, providing jobs to local citizens and forging projects to improve their quality of life. The center has gross annual revenue of $4 million and employs sixty local residents, paying them above minimum wage.

The Resource Center reflects one man's vision. Ken Dunn, a former social activist and philosophy instructor at the University of Chicago, left academia and political demonstrations in favor of "practical demonstrations" and practical politics. Though he never finished writing his dissertation, "Resources and Discontent: An Inquiry into the Interaction between the Waste of Natural and Human Resources," he made it material through real-world actions (Kalven 1991, 20). In the late 1960s Dunn began showing local merchants and street people that the cardboard boxes they burned and the aluminum cans they threw into vacant lots had value. He collected and sold materials, then offered money to people who would help him with the task. As he came to know the local poor and their socially perpetuated wasteful habits, he wanted to "connect overlooked and wasted human resource with overlooked material resource" (interview with author, Chicago, September 13, 2001).

In 1975, Dunn opened the Resource Center as a device for exploring elements of sustainable society and eliminating waste. Dunn began a recycling yard, laboring to prove that recycling works economically as well as environmentally. Today, the center manages five programs: curbside collection and transfer, a surplus warehouse for creative projects, a bicycle repair and education shop, a perishable food collection program, and an organic composting and gardening program. The sorted curbside recyclables from various housing developments are processed at an abandoned rail yard in the Garden Crossings neighborhood. It is also a destination for three hundred local "recyclable scavengers," who collect cans, bottles, and paper from dumpsters and trash cans and bring them to the center's drop-off site. When the wheels on the grocery carts used by some of the scavengers to transport valuables become worn, the center replaces the wheels so that no one will need to steal other carts.

The rail yard recycling site displays piles of sorted material and compost and a vegetable garden within spaces defined by makeshift walls. Made of carved stone remains from a demolished bank of a classical-style architecture and rusted old steel truck bodies and carcasses of VW vans from the center's old fleet, the yard is a mosaic of material, a colorful and orderly operation. The journalist Jamie Kalven describes the yard in poetic terms: "It is a strangely consoling—and even, in its way, a beautiful—place. In this setting, man-made materials take on an almost organic quality—perpetuated, reincarnated, given ongoing life by the care conferred upon them. And the postures of the workers, winnowing through these artifacts, suggest both the hard labor and the dignity of farmers bringing in the harvest" (1991, 23).

The Creative Reuse Warehouse houses material surplus and rejects donated by businesses—empty packaging, buttons, ribbons, art paper, and more—and invites artists and teachers in inner-city schools to enjoy what they could not otherwise afford. Art workshops on bookmaking, games, and even architecture using warehouse materials are conducted on some weekends. At the Blackstone Bicycle Works, a program initiated by the center in 1994, neighborhood youth can learn to repair and sell bicycles (donated by individuals and bike shops) during the summer months and after school, picking up small business and social skills that prepare them for future employment. Those who volunteer at the shop for twenty-five hours learn to fix old bikes and earn a free bike for themselves. The facility has become a local recycling company and a center that sells used bike parts and organizes bike races and special events in the neighborhood.

In another neighborhood, small pocket parks and a community garden occupy a few of the many vacant lots. It began as Dunn offered the local residents compost and lured them to organize and transform dereliction into a pleasant and productive environment. The Center's urban composting and gardening program composts yard trimmings from parks and gardens, food scraps from cafeterias, and horse manure from the stables of Chicago's mounted police. The fertilizer is then sold or donated for the community gardens program. More than 130 lots have already been cleaned and planted. Three four-acre lots on the South Side, fertilized by the center's enriched compost, are growing tomatoes that are sold to some of Chicago's finest restaurants, including that in the Ritz-Carlton Hotel. After two years of training, interns can develop farms on lots in their own neighborhoods. Plans to renovate an abandoned building into a sustainable urban agriculture center and to grow vegetables in its surrounding 8-acre land are in the making. An on-site restaurant will offer meals made of the locally grown produce. In his relentless efforts to link recycling and urban recovery, Ken Dunn envisions Chicago's estimated eighty thousand vacant lots turning into farm land, replacing the fields lost to sprawl at the urban periphery. According to Dunn, who is currently working to interest the city in his idea, the farms would provide a living for twenty thousand families.

Beyond their environmental agenda, the Resource Center, BRING, and the Wellesley dump have shaped multifaceted resources for their communities. The Resource Center has become a full-fledged social programmer, a vehicle for teaching and empowering people through sustainable living practice; BRING has become a sociable and entertaining place for creative spirits and environmental education; and the Wellesley dump has developed into an upscale community center. Designers and artists have indeed added their fair share of creative ideas for reuse and junk art installations to these sites, but because they typically express middle-upper class values and aesthetics, their own notions become less relevant in these community projects.

Exploratory Design Scenarios

Even as reuse and recycling programs are on the rise, other experiments to reduce garbage and to utilize waste efficiently are also taking place. For instance, chemists are engaged in attempts to produce edible and degradable food-packaging material. Burying garbage in ocean trenches, building ocean islands of compacted garbage bricks (and constructing power stations to produce electricity from the buried garbage), and metabolizing garbage with a genetically created bacteria that would create rich protein are other current research venues.

Meanwhile, more voices join the call for decentralization, aiming to make recycling and reuse a normal part of the everyday life and landscape. Integrated Sustainable Waste Management, an alternative approach to the centralized, conventional municipal system, advances the integration of the informal sector (the poor and small individual initiatives) with the formal (municipal and large corporations) and is being debated in professional circles. WASTE is one of several nonprofit organizations that initiates and tests pilot projects in many countries around the world.

Likewise, the architect and landscape architect Walter Hood targets the true needy, who are the most effective at making do with recyclables and reusables but are still denied access to many recycling programs. He proposes a program that explicitly accommodates the "cart people," the contemporary street scavengers, and considers inner-city neighborhood parks as potential sites for recycling collection, open every day. Each park would have a recycling structure equipped with a utility room wide enough for shopping carts to negotiate, a recycling shed, and space for children's creative projects. Residents would bring their recycling materials to the shed, and the needy could come and retrieve the items they required (fig. 1.2). In another urban project, within the confines of a freeway exit's commonly wasted land, Hood has designed a neighborhood recycling center as part of an all-encompassing complex—playing fields and courts, a community center, shops, an outdoor market area, housing, a multidenominational church, and a blues and jazz club. "Instead of city-sponsored recycling pickups, new jobs are created by turning this service over to neighborhoods" he explains (1997, 47).

Some German cities have implemented various convenient and accessible models that are well integrated into the daily life of their middle-class citizens. For instance, Munich constructed about twenty small recycling centers in the midst of suburban residential areas throughout the city. The constructions, simple but not banal, are built as a series of buildings and sheds of similar forms around a large communal courtyard filled with sculptures. The idea of integrating recycling and recreational facilities was proposed by the Dutch landscape architect Adriaan Geuze of West 8. In west Holland, Geuze juxtaposed recycling with favorite pastimes on both sides of a rectilinear dike a few miles long.

Another conceptual design proposal—*The Open Waste System Park* (1992), by the landscape architects Gina Crandell and myself—envisions a network of recycling and reuse parks interwoven throughout the city fabric (Engler and Crandell 1993). This model calls for a decentralized system of small recycling sites at three levels—the neighborhood, the central shopping center, and the major city waste redemption park. At the neighborhood level, the grassy margin between the street and the sidewalk is transformed into a practical and playful strip, made active as recyclables are frequently placed there and removed. It is equipped with roofed bins and transparent cases for collecting waste, where neighbors can sort and claim rejected goods. Areas next to neighborhood schools or parks are set aside for yard waste and the exchange and reuse of belongings. These areas, encountered by children and adults on their daily walks, become community places to socialize, play, and sort through, take apart, and reconstruct used commodities. The next level of this system is located where consumption and parking meet—the shopping mall. The parking lot houses the park and a salvage area. Convenient drive-by transparent bins accommodate outdated fashions, obsolete technology, and other used materials (fig. 4.12). The highest level of the waste park system is found at the city's central redemption facilities, which currently encompass the power plant, coal piles, the sewage treatment pools or their wetland counterparts, solid waste recovery plant, and landfills. There, buses loaded with schoolchildren and adult visitors witness the processes of nature and humans at work—the redeemed products and their applications in the landscape.

Recycling and reuse are a critical form of exchange—social, economic, and symbolic—and an expression of our relationship to the culture and material world. Over the years both market and aesthetic sensibilities have shaped places of reuse and recycling. Yet to make "waste places" truly accessible, productive, and socially and economically vital, planners and designers need to locate them strategically throughout the everyday landscape and free them of the ambience of charity or the formality of government. Designs should be simple but elegant and should, at times, forgo their tidy, elitist aesthetics and accommodate the essential presence of conflicting elements of order and untidiness, control and freedom, cleanliness and dirt. We need material recycling and reuse places where reuse and recycling intermix and where, as Walter Hood asserts, "The sacred and the profane are united, a constant reminder of their duality in our lives" (1997, 47).

Fig. 4.1. *(top)* Sorting refuge with mechanical aid, New York City, 1903. From *Scribner's,* October 1903

Fig. 4.2. *(bottom)* A billboard protesting the plan to locate a garbage transfer station in the Phillips neighborhood in Minneapolis in 1995. Photo by Mira Engler

Fig. 4.3. *(top)* "Save Waste Paper," a poster promoting frugality during World War II. National Archives, College Park, Md.

Fig. 4.4. *(bottom)* A maze of machines and conveyors at a mechanized recycling center, San Diego Materials Recovery Plant, San Marcos, Calif. Photo by Mira Engler

Fig. 4.5. *(top)* The Reusable Depot at Arcata Community Recycling Center, Arcata, Calif. Photo by Mira Engler

Fig. 4.6. *(bottom)* *World Apart,* Nancy Rubins, Washington, D.C., 1982. Courtesy Nancy Rubins / Paul Kasmin Gallery, New York

Fig. 4.7. *(top) The River of Hopes and Dreams,* a path and pavers inscribed with names and embedded with discards. Garden at the Sanitary Fill Co., San Francisco, Calif. Photo by Mira Engler

Fig. 4.8. *(bottom)* Mierle Laderman Ukeles standing near the glass bridge at *Flow City,* 1992. Photo by Daniel Dutka. Used by permission of Mierle Laderman Ukeles; courtesy Ronald Feldman Fine Arts, New York

Fig. 4.9. *(top)* View from entry bridge. The 27th Avenue Solid Waste Management Facility, Phoenix, Ariz. Photo by Mira Engler

Fig. 4.10. *(bottom)* The Book Swap at the Recycling and Disposal Facility, Wellesley, Mass. Photo by Mira Engler

Fig. 4.11. *(top)* The junkyard atmosphere at BRING Recycling, Eugene, Ore. Photo by Mira Engler

Fig. 4.12. *(bottom)* *The Open Waste System Park,* drive-by, transparent recyclable bins, and second-hand market at the shopping mall parking lot. Proposal for a network of urban parks with a common theme, waste recycling and management, Mira Engler and Gina Crandell, 1992. Used by permission of Mira Engler and Gina Crandell

Fig. 5.1. *(top)* Formal and grand geometry of structures and pools characterizes the modern sewage plant. A bird's-eye view of the Water Pollution Control Facility in Ames, Iowa. Photo by Mira Engler

Fig. 5.2. *(bottom)* A sewerman ascends with a box of rats. Paris, 1911. Photo by Boyer. Courtesy Collection of Roger-Voillet, Paris

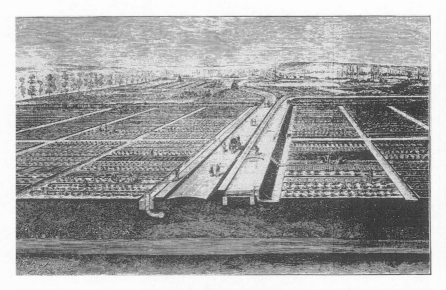

Fig. 5.3. *(top)* Sewage farming at Gennevilliers in the 1870s. Courtesy L'Illustration, Paris

Fig. 5.4. *(bottom)* Paris' underground sewers below and tree lined boulevards above, designed by engineer Baron Haussmann and landscape designer Adolphe Alphand, 1860s. From Adolphe Alphand, *Les Promedanes de Paris* (Paris: J. Rothschild, 1867–73)

Fig. 5.5. *(top)* Boat trip through the sewers, Paris, 1896. Courtesy Collection of Roger-Voillet, Paris

Fig. 5.6. *(bottom)* Wards Island Sewage Plant, designed in the Art Deco style during the New Deal era, 1937. Used by permission of New York City Department of Environmental Protection

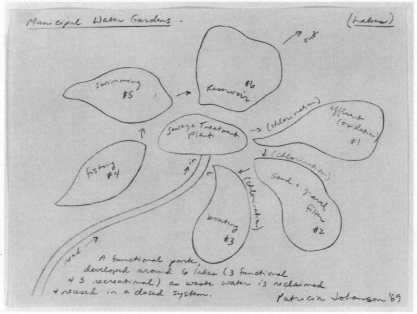

Fig. 5.7. *(top) Artist's Shit,* Piero Manzoni, 1961. © 2002 Artists Rights Society (ARS), New York / SIAE, Rome

Fig. 5.8. *(bottom) Municipal Water Gardens,* Patricia Johanson 1969. Used by permission of Patricia Johanson

Fig. 5.9. *Tides Out/Table Set,* Buster Simpson, 1984. Used by permission of Buster Simpson

Fig. 5.10. *(top)* Gut Marienhof Sewage Plant, Munich, Germany, 1990s. Soaring structures and bold design. Architect: the office of Kurt Ackerman, Munich. Photo by Ingrid Voth-Amslinger. Used by permission of Kurt Ackerman

Fig. 5.11. *(bottom)* Oceanside Water Pollution Control Facility, Fort Funson, San Francisco, Calif., is buried in a concrete canyon next to Highway 1. Photo by Mira Engler

Fig. 5.12. *(top)* West Point Wastewater Sewage Treatment Plant, Seattle, Wash., is screened with a series of planted terraces. Photo by Mira Engler

Fig. 5.13. *(bottom)* An early proposal for the North River sewage plant by the architect Phillip Johnson, 1967. New York City Department of Environmental Protection

Fig. 5.14. *(top)* North River sewage plant, New York City, is covered with park and recreational facilities, 1989. New York City Department of Environmental Protection

Fig. 5.15. *(bottom)* Donald C. Tillman Wastewater Reclamation Plant, Los Angeles, Calif., 1995, is adjoined by the Japanese garden and administration building on the other side of the fence. Used by permission of City of Los Angeles Department of Public Works

Fig. 5.16. *(top and middle)* Lighting design concept drawings for the Newtown Creek Water Pollution Control Facility, Brooklyn, N.Y., by L'observatoire International, 1999. Used by permission of L'observatoire International and New York City Department of Environmental Protection

Fig. 5.17. *(bottom)* *Project for the Periphery,* Newtown Creek Water Pollution Control Facility, Brooklyn, N.Y., Acconci Studio, 1998. Used by permission of Acconci Studio and New York City Department of Environmental Protection

Fig. 5.18. *(top)* The Interpretation Center and marsh, Columbia Boulevard Wastewater Treatment Plant, Portland, Ore. Architect: Miller Hull Partnership and Sera Architects. Landscape architect: Murase Associates. Photo by Scott Murase. Used by permission of Murase Associates

Fig. 5.19. *(bottom)* The Marine Park Wastewater Reclamation Facility in Vancouver, Wash., encloses its facilities with structures and integrates the Water Resource Education Center and a park, 1997. Used by permission of American Public Works Association, Kansas City, Mo.

Fig. 5.20. *(top)* "Buble," from Su-Chen Hung's *Water Spells,* installation at the aeration basins in the 23rd Avenue Wastewater facility, Phoenix, Ariz., 2000. Photo by Mira Engler. Used by permission of Su-Chen Hung and the City of Phoenix Arts Commission

Fig. 5.21. *(bottom) Waterworks Garden,* Renton, Wash., Lorna Jordan, 1996. The East Division Wastewater Reclamation Plant tanks are juxtaposed against the marsh of the garden. Photo by Mira Engler

Fig. 5.22. *(top)* The Living Machine in the Findhorn Foundation, an international center of spiritual education and personal transformation in Scotland. This machine treats the wastewater for 350 people at a capacity of 17,000 gallons per day. Built in 1995 by Ocean Arks International. Used by permission of Ocean Arks International, Burlington, Vt.

Fig. 5.23. *(bottom)* Recalling geometric garden beds, this wastewater treatment plant in Gorgonzola, Italy, utilizes duckweed to clean sewage. Viet Ngo, Lemna Corporation, 1989. Used by permission of Lemna Corporation

CHAPTER 5

Sewage Treatment Plants
Wastewater Gardens

Do we really know the city if we do not know its sewer? Not according to Victor Hugo, who writes in his 1862 novel *Les Misérables:* "The history of mankind is reflected in the history of cloaca. . . . The sewer of Paris was a formidable ancient thing, both sepulchre and refuge. Crime, intelligence, social protest, liberty of conscience, thought and theft, everything that human laws pursue or have been pursued has been hidden in it" (1998, 1064).

The Multivalent Sewer
The ditch of truth

An educated discourse about the sewer necessarily starts with Hugo, for whom the sewer was a central, most potent image of society during France's First and Second Empires. At the climax of *Les Misérables,* Hugo opens the city sewer to his readers, making it the setting of the drama. The protagonist, Jean Valjean, saves the young revolutionary Marius as he carries him unconscious from the barricade and counterrevolutionary squads, taking him through the underworld of Paris to safety. The sewer is portrayed as a site of refuge, though a long introductory narrative paints it as a multifaceted, multivalent entity.

Hugo's sewer is real and imagined, physical and symbolic, debased and wholesome, repulsive and fascinating. It is an asylum and a trap, a place of resistance and submission. It is a dwelling for criminals, revolutionaries, and the poor. It is a place of surprises, of decomposing bodies and faded, lost treasures. Above all, it is a cultural symbol of moral disintegration and political disorder, a resting place of all failures, a palimpsest that displays "the remains of every cataclysm" and every misdeed. Thus, it becomes a laboratory for the social observer and historian and a site of social discourse. Although it is repressed, it remains a sincere realm, *a ditch of truth:* "the conscience of the town where all things converge and clash. There is darkness here, but no secrets. Everything has its true or at least its definitive form. . . . Every foulness of civilization fallen into disuse, sinks into that ditch of truth wherein ends the huge social downslide, to be swallowed, but to spread. It is a vast confusion. No false appearance, no white-washing, is possible; . . . The last veil is stripped away" (Hugo 1998, 1065).

A true reflection of the world above, the sewer is however not a passive receptacle. Rather, "A sewer is a cynic. It says everything. . . . This sincerity of filth pleases us and soothes the spirit" (1998, 1065). The nineteenth-century sewer circulating below the surface of the Parisian *mentalité* was a locus of relief from the dishonest aboveground world.

The mental image

The French have always reserved a special place for the sewer (and for excrement) in their urban history and cultural imagination.[1] Hugo's potent work taped into and shaped this mental image, employing metaphor, simile, association, and personification. He writes of a "wretched vomit" of the sewers into rivers and calls the sewer a strange feat, the "abjectness of grandeur," a "colossal sink," the "disease of Paris," and the "evil of the city blood." He likens its outfalls to "dragon mouths, breathing hell upon men" and to "forest caves." Hugo sketches the maze of the sewer as a vast sea-plant or a tree—the main sewer is a tree trunk, its lesser channels are the branches, and its dead ends the twigs. "If one thinks of Paris lifted up like a lid, the view of the sewers from above would resemble a great tree-trunk grafted onto the river" (1998, 1064). Alternatively, he analogizes it to "a grotesque jumble of eastern letters attached to each other haphazard, by their sides or their extremities" (1064).He also equates the sewer to human organs of arteries, stomach, and intestines and personifies it, giving it a mouth with which to vomit and an arm with which to strike back.

People conjure up images of the sewer that derive from biblical and other literary sources. These can refer to both security and damnation: the prophet Jonah rode to safety in the intestine of Leviathan, whereas sinners in Dante's eighth circle of Hell wallow deep in excrement flowing from human privies (1954, 197). Contemporary cartoons and films still portray the sewer as a home and hiding place for good and for evil. *Teenage Mutant Ninja Turtles* emerge from the sewer to save the world (when they do not watch TV and order pizzas), and in the film *Batman Returns* (1992) the sewer serves as a lair for the vile Penguin.

The sewer is an essential ingredient of the urban mentality. It embodies the city's shadow twin, its base receptacle. The renowned Italian writer Italo Calvino captures and draws on the city's two self-projections—the celestial and infernal—in *Invisible Cities*. He describes the city Beersheba, whose

> inhabitants honor everything that suggest for them the celestial city: they accumulate noble metals and rare stones. . . . They also believe, these inhabitants, that another Beersheba exists underground, the receptacle of everything base and unworthy that happened to them, and it is their constant care to erase from the visible Beersheba every tie or resemblance to the lower twin. . . . [T]hey imagine that the underground city has overturned rubbish bins. . . . Or even that its substance is dark and malleable and thick, like the pitch that pours down from the sewers, prolonging the route of the human bowels, from black hole to black hole. (1974, 111–12)

Calvino calls Beersheba an empty vessel, a greedy city and one that "only when it shits, is not miserly, calculating, greedy" (111). The act of shitting, therefore, entails the only positive, candid moment.

Undoubtedly, the dominant mental image of the sewer is that of an underworld laden with repulsive and diabolic references—miasma and pitfalls, fetid odor, obscurity, ominous shadows of criminals, slime, and excrement. Since its full development in the nineteenth century, the sewer has built on and elaborated a symbolic system that associates the low with the dirty and has maintained a tenacious social stigma.

The sociological terrain

The sewer has been generally regarded as an immoral site, but Hugo used the sewer to excavate the social narrative of history, to unearth submerged groups and sociopolitical oppressions and failures. Likewise, in his compelling 1991 book *Paris Sewers and Sewermen,* the social historian Donald Reid recounts and interprets social practices and representations through the sewer in order to contest traditional urban and labor history. He argues that the sewer offers a deeper, more truthful terrain on which to study social history and politics. According to Reid, the historian searching for events finds them out in the open, etched on exterior surfaces—the battles, revolutions, royals, and statesmen—but the historian searching for morals finds them in undergrounds, hidden in dark interiors, among the poor, orphans, the infamous, and unfortunates.

The sewer is the embodiment of the state, and yet antisocial elements of society and challengers to the social order often found refuge in the sewer. Its crypt-like, hard-to-supervise passageways provided a safe haven for criminals, revolutionaries, and victims alike. Sewers have always been home to an underworld of crime. Criminals used the sewers to break into and rob banks and to escape police. Prisoners attempted escape via the sewers. The sewer was a place from which to undermine a ruling power. Rebels, including the famous French revolutionary and journalist Jean-Paul Marat, hid in the sewers for extended periods. During World War II the French Resistance made use of Paris sewers and briefly located its headquarters and first aid station there. The Jews of the Warsaw ghetto in Poland fighting against the Nazis operated from the sewers until they were found and eradicated.

Yet the sewer is not only an extension of the criminal and antipolitical underworld; it is equally a resting place for the relics of the misdeeds of the city's powerful and well-to-do, a locus of repression of the powerless, and the haunts of what is considered impure and immoral. The dirty and foul, the poor and the uncivilized, the slum and the ragpicker have always been lumped together with the sewer. While the sewer drained the higher ground of the rich, it overflowed the lowlands of the poor. It may be no surprise that the novelist and caricaturist Émile Zola focused on human excrement to point out social ills and the failures of those in power.

Also, the sewer has always been gendered and equated with prostitution and the female sex organ. Some powerful health and social apostles have conceived of the sewer as a feminine site controlled by erratic nature and deprived of reason. The French public health expert Alexandre-Jean-Baptiste Parent-Duchâtelet, who made a career studying the Parisian sewers, made this analogy in his famous early-nineteenth-century book on prostitution in Paris (Parent-Duchâtelet 1837). The English sanitary reformer Edwin Chadwick lumped together as immoral the sewer, prostitution, slums, and disease in his well-regarded writings on unsanitary disorder among London's laboring poor.[2] But the sewer has also offered a last resort and emergency dwelling for the poor. Even today in third world countries, the homeless—for example, poor children in Ulan Bator, the capital of Mongolia—reluctantly adopt the sewers as their dwelling during the cold winter months.

The ideological map

The sewer represents an evolving and paradoxical social construct. The development of industrial technology over the past three centuries has largely shaped the shifting societal conceptions of the underground world. Rosalind Williams's intriguing book *Notes on the Underground* suggests that it was late-seventeenth-century mining operations that opened up what had been considered formidable devil's grounds to new, less fearful associations. As perceptions changed and what had been ugly and disgusting began to appear sublime, people felt a combination of awe and anxiety.

In the mid-nineteenth century, increased civic order and technological advances in underground utility and transportation systems brought new reactions, and sublimity was eventually replaced with fantasy and wonder (Williams 1990, 83). Cities such as Paris and London provided inhabitants a safe place to gaze into the newly constructed, clean, and tamed underworld. Sewers became a spectacle of enlightenment. The sewers of Paris were a respected item in the Baedecker tourist guide (as discussed below). In *The Politics and Poetics of Transgression*, Peter Stallybrass and Allon White (1986) contend that as the city was planned so that the dirty was made invisible to the bourgeoisie, the self-assured bourgeois actively sought out filth and was attracted to the newly hidden, subversive domain. Filth (moral and physical) represented the very antithesis of gentility, fearful and fascinating at once.

As sewers continued to become more efficient and sanitary techniques proved effective, both anxiety and wonder faded. The need to encourage public support of sewer projects and public faith in the new technology diminished, and so did interest in visiting the sewers. The tamed, modern sewer of the twentieth century, out of sight and out of mind, no longer provided novel entertainment; it was slowly stripped of its earlier diabolic as well as romantic associations. It became a sterile, efficient, and functional engineering apparatus. Since the later decades of the twentieth century, however, the sewer has become a "second nature," an unpredictable landscape in which disaster lurks (Williams 1990). Consequently, a century after it was modernized, as the sewer's effect on the greater environment proved detrimental, sewers and sewage plants were again opened to the public gaze. Regaining public support for further construction is a decisive factor in this new development. The modern, sanitary sewer—essentially a technological creation—is painted as a necessary evil, a volatile "neighbor" that demands understanding and care if a healthy relationship is to be maintained.

The technological landscape

A product of the city and a result of dense urban population, the sewer was one of several scientific tactics for fixing the city. Seeking to improve social and physical order, sanitary and civic engineers of the late nineteenth century were optimistic that technology could address urban ills and solve its sanitary practice. The sewer was perceived as a physiological apparatus, literally (and metaphorically) modeled after the human body's system for processing waste. Conduits of gradually increasing size convey waste particles in water from every urban "cell" to a point of discharge—the outfall. On route, they pass through a mechanical cleansing "kidney," a treatment plant located near a body of water. The built waterways connect all city establishments, including the most private and intimate domestic receptacles—toilets, sinks, and tubs—to natural waterways. The tireless system serves a constant flow of murky fluids; peak volumes occur before and after the daily work hours, as well as during television commercial breaks during popular programs.

The sewage system has been built slowly and incrementally, tested and refined on a trial-and-error basis. It started as a simple conduit that was meant to remove sensory irritants by draining city streets, and it grew into a centrally controlled, sophisticated, and capital-intensive transport and cleansing apparatus. The early sewers were open street gutters or small, closed earthen pipes not flushed by water. They often leaked, clogged, and fouled the air. Improvements in pipes and construction techniques and the use of water and gravity as agents of flow resolved many of these problems. Sewers and waterworks became a continuous, circular system, mirror images of one another. The potable water treatment plant and the wastewater treatment plant anchor the two opposite points of the cycle; the first cleans water from a natural source for use, and the second cleanses wastewater before discharging it back to its origin. In between, the water is dispersed into increasingly smaller and branching pipes and returns as wastewater via increasingly larger pipes.

Early cities integrated the stormwater and wastewater systems into the same network, but the combined overflow became a serious source of pollution. During heavy rainfall, wastewater bypassed the plant, backing up into the street and running directly into local waterways. Since the early 1900s, combined systems have been slowly replaced with separate systems in order to prevent the troubling pollution of untreated overflow.

A network of pipes—laterals, submains, and interceptors—and pumping stations collect and convey sewage to a point of treatment. Laterals are the smallest pipes in the network, usually no less than 8 inches (200 millimeters) in diameter, and carry sewage by gravity into larger submains or collector sewers. Collector sewers tie into a main interceptor or trunk line, which is built of reinforced precast concrete pipe sections up to 15 feet (or 4 meters) in diameter, and connect to a treatment plant. Special cameras designed to travel inside sewer lines are used to monitor new construction and to detect flaws before and during usage. Occasionally, smoke bombs are ignited to test for cracks. Pumping stations are built when sewage must be raised from a low point to a high point, or where the topography prevents downhill gravity flow. The structures housing the pumps are called "lift stations."

To ease the water flow at intersecting lines, a cylindrical catch basin is used; it often connects with a checkpoint manhole. These manholes (and the end outfall) are the only points of access into the system. Manholes, the sole tangible manifestation of the world below, have a special place in the human imagination. An open manhole is often portrayed in caricatures and films as a formidable trap or a chancy escape route. Sewers did not initially have manholes for fear of gas escape; pipes had to be broken and repaired when access was needed. Trade publications begin to show them in the 1860s and more often 1870s. A manhole, made of precast concrete, is at least 22 inches in diameter and 6 to 12 feet deep. Its cover is "married" to its frame and must be tightly seated to avoid rattling when vehicles pass over. The 250- to 300-pound metal covers are generally circular, to ensure easy fit and to facilitate their manual movement; they are textured to provide traction for the traffic above.

Embellished with unique patterns—a kaleidoscopic assortment of knobs, ridges, circles, slots, crosses, diamonds, flowers, scallops, and stars—and often bearing the name of the company or utility whose pipes they cover, manhole covers are beginning to be seen as a form of industrial art. According to Mimi Melnick, who has studied and documented hundreds of manhole covers throughout the United States, "They are veritable archaeological gems, embodying the rich design vocabulary of bygone eras;" (1994, 28). In June 1995 the City Council of Los Angeles declared the local manhole covers historic artifacts, thus protecting them during street maintenance and remodeling.

At the end point in the sewer system, a cleansing facility treats and purifies wastewater, guarding water quality and public health. The "influent," the sewage flow that enters a treatment facility, is turned into an "effluent" containing mostly water and various solubles and bacteria. In professional parlance, the sewage plant removes suspended particulates and biodegradable organics, destroys pathogenic bacteria, reduces nitrates and phosphates, and neutralizes industrial wastes and toxic chemicals. Literally, it separates water from waste and detoxifies the waste.

The sewage plant comprises buildings, structures, machines, containers, tanks, and ponds or lagoons, resembling from the air a grand formal garden (fig. 5.1). Whether simple or complex, small or large, it commonly combines mechanical, biological, and chemical processes. As water moves from one part of the facility to another, it is subject to various mechanisms, as the waste is pumped, screened, deposited, and digested by aerobic and anaerobic bacteria. The anatomy of this technological kidney constitutes an intricate maze that is easier to represent with a diagram than to describe with words. Sewage enters the plant via an influent pump. Past the initial screen bars and grit chambers, where most solid suspects (including rocks, condoms, and wedding rings) are captured and removed, there are pumps, sedimentation tanks that act as primary clarifiers, scrapers, aeration tanks that are the secondary clarifiers, final clarifiers, and chlorine contact basins or ultraviolet light for final disinfection. The complete cycle of cleansing takes ten hours to three days. The process is divided into levels of treatment—primary, secondary, and tertiary (or advanced)—and the level required varies, depending on local environmental conditions and governmental standards.[3] The flow of water inside and between the tanks is an engineering feat and a visual spectacle. Sometimes a great gushing waterfall is formed at the effluent weirs or outfall where clean water leaves the plant. At other times, water is carried away in pipes and used for irrigating crops, replenishing aquifers, and filling recreational lakes.

Sludge follows its own course and timeline (between fourteen and thirty days) through processes of digestion, thickening tanks, and dewatering. Sewage dewatering facilities have only recently begun appearing on plant grounds, as bans on sludge disposal in the ocean and in dumps came into effect in 1992. The tall, massive buildings house machines that spin dry the sludge with high-volume centrifuges, disinfect and chemically "polish" gases, and produce a semisolid black residue called "cake." The cake is loaded on trucks or vessels that dock at the plant's own sludge port. It is then used for various land applications, such as inactive landfills, parkland, and golf courses, or it is made into a fertilizer and dry pellets that are burned to produce energy. The notoriously smelly beds that used to dry sludge in the open have been mostly eliminated, and with them the tomato plants that sprouted and flourished at their edges.

The mechanical control room, the plant's brain, is the most informative and fascinating place to visit. It is connected to all the parts of the system, constantly measuring and monitoring their processes via an electronic "nerve system." Its walls are covered with illuminated panels with switches, buttons, and gauges superimposed on the plant scheme. Computers and display screens analyze data and project images from various plant functions. Organized tours usually begin in the control room then follow the wastewater through the maze of cleansing stages and technological wonders.

Sewer habitats

Despite their links to advanced technology, sewers and drains are a favorite habitat for animals. The sewer rat, the infamous and hated brown rat, is a persistent adjunct. It is another despised irritant, a symbol of what we wish to forget; it provides evidence of our limited ability to suppress and hermetically close off the sewer (fig. 5.2). When rats escape from the enclosed line, they transgress private boundaries and disclose the sewer's pervasive underground presence.[4] Small-town sewers continue to serve as a secure habitat for rats, raccoons, and opossums. These rodents appear briefly above ground, prey on residential vegetable gardens and trashcans, and disappear underground.

Sewage plants, too, provide an impressive ecological habitat. Even well-built plants house thriving populations of insects, rodents, waterfowl, and fish. The open tanks are breeding grounds for mosquitoes, which are welcome prey for birds, who appear in striking variety. Seagulls, pigeons, and pheasants eat organic solids in the sewage; barn swallows eat mosquitoes; hawks and owls prey on the birds and rodents. Fish that occur naturally in the flow, such as carp, bass, goldfish, and catfish, thrive in the water tanks, as do crayfish; other fish, such as guppies and gambusias, are introduced into the final tanks to help eat mosquito larvae and other organisms in the water. The mosquito fish, as it is called, dwells near the surface and enjoys the relatively warmer temperatures of the wastewater tanks. Rats, snakes, and rabbits can also be detected in some "less sanitary" sewage plants, and spiders thrive in the darker corners of older plants, enjoying an abundant supply of mosquitoes.

The organism in greatest supply (and in constant labor) is the bacterium, an integral part of every plant. Bacteria and other microorganisms are responsible for breaking down the sewage sludge in both the aeration and digestion tanks. At the secondary clarifiers, or aeration basins, bacteria break sludge down in the presence of oxygen; anaerobic bacteria consume the organic material in the sludge, producing carbon dioxide and methane gas in the process. The gas is either burned or captured and used to generate electricity. Newer plants, especially those whose clarifiers and pools are completely enclosed are obviously tidier, more sanitized, and lacking in wildlife.

The power of the sewer to evoke multivalent, contradictory sensations resides in its multifaceted ecology, technology, ideology, sociology, and imagery. However, it owes much of its aura to the very content that travels in its guts, to human muck. Without it, the sewer is simply a pipe full of water.

Sewage Idiosyncrasies

Sewer muck

Sewers carry whatever is small enough to go down a drain. Because they are pervasive, easy to use, and hard to monitor, in them can be found matter both benign and toxic, valueless and valuable, profane and sacred. Sewermen used to recover valuables from the murky fluids. Reid tells us that the nineteenth-century Parisian sewer administration dredged and salvaged goods from the newly constructed sewers and sold them to junk dealers (1991, 53). Laborers working in the great sewers of ancient Rome found a female statue and made her the goddess of the sewer, Cloacina. They bestowed on her the power of protecting them from the evils of the sewer. Titus Tatius had a statue of Cloacina in his magnificent privy, which was built as a shrine to her glory (Horan 1966, 17).

Sewer water, or sluice, is peculiar, a malady and a cure at once. Known for its disease-ridden and dangerous emanation, the sewer is considered havoc. Many sewermen found deadly sickness in its guts, yet others attested to its healing power and the liquid's ability to cure wounds and immunize against diseases to which other urbanites were susceptible. Parent-Duchâtelet, who ventured into the foulest sections of Paris's sewer to study the effects of its vapors, argued for the therapeutic qualities of excrement and sewage (B. Miller 2000, 33). Laboring to defy what he saw as the prevailing misconceptions about sewage, Parent-Duchâtelet found that people who immersed themselves in the sewage basins of Montfaucon were cured of rheumatism and other infirmities (Corbin 1986, 212). The special chemistry of sewage sluice was said to have other beneficial qualities. According to Reid, the sluice smoothened and softened the Parisian sewermen's boots, making the leather supple, tough, and beautiful, and highly sought after by ladies' bootmakers (1991, 53).

The content of the matter disposed of in the sewer changed over the centuries. The first sewers drained only surface water and marshy areas. Some public toilets were connected to the cloaca in ancient Rome, but private vaults and cesspools were connected to primitive sewers only in the eighteenth century, and then often illegally. Stormwater and residential sewage were directed to the modern nineteenth-century sewers. Industrial discharge was added to the mix in the early twentieth century, turning sewage into a liquid saturated with toxins and even radioactive waste (aided by household cleaners). The in-sink garbage disposal added its fair share of nitrogen and the washing machine its phosphates, two nutrients essential for plant growth though damaging in large quantities. Some consider sewage to be fundamentally useful matter, which is mixed too plentifully with harmful substances and directed to the wrong place.

Waste professionals speak of the source, type, and concentration of pollutants. They may originate from a "point source" (from a single directed pipeline or channel) or from a "dispersed source" (a broad, unconfined area from which they make their own way to a body of water), which is hard to control. Sewage is of three types—domestic, industrial, and stormwater. Today's sewer muck (or sewage) contains an impressive list of "water pollutants"—pathogenic organisms, oxygen-demanding wastes, plant nutrients, synthetic organic chemicals, inorganic chemicals, sediments, radioactive substances, oil, and heat. In the past, however, sewage was used as a beneficial fertilizer.

Sewage farming

Sewage works dealt a blow to the long-standing agricultural use of human excreta—the direct application of human waste to fields. Although societies have often been interested in the fertilizing virtue of sewage and thus its monetary value, the practice has been dropped and revived at various points throughout Western history. Following a few centuries of neglect, sixteenth-century Europe rediscovered it. When cesspools were mandated in large cities, their contents were taken to fields. In England and America, urban scavengers collected privy matter and horse manure and transported it at night to fields in the country (hence the name "night soil"). In late-eighteenth-century Paris, the collected fecal matter was delivered to the sewage dump in Montfaucon and dried. The resultant poudrette was warm and ready for combustion.[5]

As nineteenth-century sewers began sending human manure directly into rivers, manure merchants, chemists, and farmers loyal to filth's therapeutic properties loudly protested the wastefulness of removing the substance from human life. Cleaning companies demonstrated against the apparatus that was about to replace them. A fierce cultural debate regarding sewer construction ensued. The same sanitary, efficient apparatus that drew the bourgeoisie's awe and admiration became a subject of criticism. Health experts and social reformers argued that sewers were faulty in essence. They blamed the underground pipes for impoverishing the land and its people, depriving the soil of fertility, wasting a valuable resource (human manure), denying people subsistence, fouling their waterways and air, and bringing about disease. They advocated the long-lost agrarian practice of turning dung into productive fertilizer through sewage farming.

The idea of sewage farms offered not only a practical solution but also the sociopolitical message of a never-ending cycle. Edwin Chadwick spoke of the Egyptian symbol of eternity, of "bringing the serpent's tail into the serpent's mouth" (Melosi 2000, 43); and the French socialist Pierre Leroux came up in 1834 with his theory of "circulus," advocating irrigating and fertilizing the fields with the "liquid manure" of the city (Leroux 1853). Chadwick proposed to transport sewage in enclosed pipes to agricultural fields in order to enhance both the land and human living conditions in the city, as well as to pay for the expensive mechanical system. In the 1870s Leroux connected the social malpractice of sludge management and economic impoverishment, going so far as proposing that people in the city gather their own dung and give it as payment to the tax collector. Influenced by Leroux's ideas, Victor Hugo joined the criticism, calling the sewer a mistake that would have serious social and economic repercussions. He noted that France alone disposed half a billion francs worth of human manure per year, mindlessly spewing it via the rivers into the Atlantic (1998, 1062). Other social visionaries, including Karl Marx, also considered sewage farming an avenue to a better tomorrow, a new political and biological order; Jules Verne incorporated filtration and sewage farming into a model city and commune, presenting them as socioeconomic therapies (Reid, 1991, 55).

Throughout the nineteenth century, experiments in the social and environmental benefits of sewage farming were conducted in England, Scotland, France, Italy, Germany, and even the United States (Melosi 2000, 214). In the early 1800s, several towns in England and Scotland, including the oldest and most successful farm near Edinburgh, began irrigating their fields with sewage. In 1880, nineteen towns disposed of their sewage directly in the fields.[6] In America, Lenox, Massachusetts, used its sewage for agriculture. By the end of the century, twenty-four other American municipalities had joined in (Melosi 2000, 214). In France, model farms in Clichy (1869) and Gennevilliers (1872) were a great success (fig. 5.3). They transformed the land, supported farming families, and supplied produce to the city. Donald Reid tells us of visitors touring the farms on Sundays and witnessing the beneficial use of their own sewage in the previously dry and infertile Achères region. They rode out 10 kilometers on the small train that was used the rest of the week to transport produce to the city in order to see these *"luxurious, odorless grounds. . . .* At the behest of the guided tour, visitors drank a glass of the limpid, filtrated water, and could compare its appearance with the sewage-laden liquid they had seen flowing into the fields" (Reid 1991, 66).

Reid claims that "Sewage farming incorporated the duality which characterized the sewers themselves: at once a technical response to a problem of urban waste disposal and a stimulant to the social imagination" (1991, 69). Sewage farming faded in the early century as an ever-increasing volume of sewage presented maintenance problems and made the process too land- and labor-intensive. Artificial screening and chemical scientific procedures took over, and sewerage was separated from farming, treated purely as waste. Beginning around 1890, intermittent filtration, a method using sand and soil as a filter medium, came into wide use. Still later, these earth-based methods were replaced with biological processes and built structures (Tarr and Dupuy 1988, 170). They soon became the subject of criticism once again, as environmentalists blamed sewer technology for aggressively intruding on the environment, fouling waterways, and undermining the economy.

Sludge, left out of the sewage treatment process, was dumped onto land and into waters. But a century after the utility of human manure was forgotten, attitudes have come full circle. Shit has become profitable again. The EPA has reclassified sludge from a hazard to fertilizer, officially calling it "biosolids." More than a quarter of the sewage sludge in the United States is now applied to farmland, but the debate remains as forceful as ever. Opponents warn us of unknown consequences, arguing that what goes down the drain and into the sewer is unpredictable and hard to monitor (Rockefeller 1996). They advocate treating sludge as a toxic waste and call for a policy that will prevent the production and accumulation of sludge altogether. Until risks are better understood, sludge might best be applied to nonagricultural land, especially to already polluted and damaged land such as landfills and quarries. Indeed, for the last three decades sludge has been used for "bio-remediation"—that is, restoring degraded lands, such as former strip-mines.[7]

Dried up and bacteria free, dewatered sludge is deprived of the smell that is so deeply ingrained in the idea of sewage.

Sewer odor

Sewage stinks. Smell has been its persistent companion. Like the rat, it also escapes the enclosed line and discloses the sewer's pervasive underground presence, but it is most intolerable for transgressing class boundaries and intruding on personal space. Prior to the construction of modern sewers, human feces conditioned the experience and colored impressions of the city.

People used mysterious and alarming methods of deodorizing—from fumigation using salt and sulfuric acid, to paving, drainage and ventilation, to musk and eau de cologne (Corbin 1986, 92). These scents gradually lowered the threshold of tolerance for everything offensive in the muted olfactory environment. Dominique Laporte claims that perfume (*pare-fumier*) has always existed in contrast to manure and exposed the fragile boundary separating good smells from fetid ones. He tells us that cesspools were aromatized in the late eighteenth century with bergamot, orange and lemon essence, distilled lavender, orange blossom essence, cloves, and countless other essences and oils (2000, 80).

As human waste was increasingly collected in sewers, sewer gas or miasma, the smelly result of the fermentation of the organic muck, was targeted as the source of epidemics and attacked by health experts and sanitarians until Pasteur's theories were clearly understood. That sewer workers sometimes died in the sewers only reinforced this belief. As already noted, the world's most famous sewer advocate, Parent-Duchâtelet set out in the 1820s to prove wrong those who linked the sewer smell to death, but he was unsuccessful. In the late nineteenth century, contact with sewage and excrement continued to be discouraged for fear of the miasma they emitted (Tarr 1999). Such fears made many people reluctant to introduce a water closet inside their house; thus, the fixture was often installed outside. Even today sewer fumes are considered dangerous; workers enter sewers in pairs, wearing masks and safety harnesses. The National Safety Council has declared sewer work one of the most hazardous occupations in the country (Melnick 1994, 16).

The study of smell and the development of new deodorizing tactics were long considered frivolous or irrelevant, but they have recently begun to receive considerable attention. Nowadays, smell determines real estate value and provides ammunition for public opposition to the siting of sewage plants (and other smell-generating environments, such as hog farms). Odor control became the focus a multimillion-dollar research projects, which produced the technology necessary to win approval for constructing new sewage plants. Odor-producing areas, sometimes entire lagoons, are being enclosed in special structures. In some sewage plants gigantic fans located next to clarifiers pull air away from residential areas. Foul air from certain plant functions is channeled into large tanks where it passes through odor-absorbing chemical scrubbers and activated carbon filters before the vapors rise through smokestacks to the air. "Electronic noses" in the form of computer sensors that recognize bad and good smells are monitoring pig farms, chemical factories, and sewers. Analysis of the odors is provided in digitized form; without ever being sniffed by a human nose (Klein 1995, 19).

Laboratories for the study of smell are being built inside sewage plants. In the Achères sewage plant near Paris, an area with a long history of sewage dumps and sewage farms, a new environmental observatory (in fact, a smell laboratory) traces and researches bad smells with the goal of remedying them. A visitor center is integrated into the laboratory, and the public is invited to participate in surveys and evaluate smells, providing a kind of jury on odors. The architects Yves Brangier

and Philippe Primard located the building at the far end of the old drying beds, and its patio and main hall open out toward the emptiness of
that vast area (Primard and Brangier 1995, 47).

Language deodorized

Like smell itself, the English language was gradually deodorized. In the
twelfth century, "smell" was an inclusive term for any olfactory property, whether agreeable or otherwise—perfume, aroma, stench, stink,
and scent.[8] In the fourteenth century the word "odour" (or odor) was invented. It initially referred to a sweet or pleasant scent, whereas "smell"
became mostly identified with the unpleasant. "Odor" was frequently
used in the seventeenth century and is almost solely used today in public and professional discourse in the United States.

The etymology of the words "sewer" and "sewage" reflect changes in
technology, perceptions, and moral judgment. In early agrarian society
there was no such thing as sewage. The Latin *cloaca* originally referred
to the sewers of ancient Rome and later was used in biology (describing
a common chamber and outlet in some vertebrates into which the in-
testinal, urinary, and genital tracts open). "Sewer," which dates to the fif-
teenth century, was initially used to describe "an artificial watercourse
for draining marshy land and carrying off surface water into a river or a
sea." By the seventeenth century, however, it referred to a covered, usually underground channel or conduit carrying away surface water.
"Works of sewers" were any works of defense against inundation from
the sea, including seawalls, ditches, embankments, gutters, and so on.
"Commissioners of sewers" were persons designated as responsible for
repairing and maintaining such works.

The French adopted "cloaca" to describe their early urban drains. The
terms "sewer" and "cloaca" (in French, *égout* and *cloaque*) were used in-
terchangeably during the early nineteenth century. Only later was a clear
distinction drawn between the two: *égout* referred to a conduit equipped
with an outflow and washed with water, and *cloaque* described a stagnant
and putrid channel (Reid 1991, 35). *Cloaque* soon became associated
with disease and was abandoned. As the sewer came under control and
was more rigorously segregated from and less visible to the city above, it
gained status and was linguistically reformed, absorbed into a discourse
of respectability. "The Sewer today has a certain official aspect. . . . Words
referring to it in administrative language are lofty and dignified. What

was once called a sluice is now a gallery, and a hole has become a clearing" (Hugo 1998, 1072).

In nineteenth-century America, as human feces became a subject of concern and a topic of discourse, the word "sewerage," meaning "the removal or disposal of sewage via an artificial, usually a subterranean conduit," was used along with composite terms, such as "sewage carriage system" and "water carriage sewerage." The terms "sewage" and "sewer" dominated the scene for most of the twentieth century. In the 1970s, however, "effluent" and "wastewater" took over. "Effluent," derived from "flow" or "outflow," was used to refer to fluid waste from industrial works beginning in the 1930s. "Wastewater" implies discarded and useless water eliminated from the useful life cycle. The two words, environmentally correct expressions that lacked the sense of contamination attached to previous terms, replaced the word "sewage," which has almost disappeared from the professional vocabulary. Likewise, the "sewage treatment plant" was renamed "wastewater recovery plant," "wastewater reclamation center," "wastewater recycling plants," or alternatively "water pollution control facility" and "water reclamation facility." The latter two, today's most popular, got rid of the word "waste" altogether. Indeed, a refined vocabulary of cleanliness emerged as an ethical discourse, driving all unhealthy humors from the genteel lexicon.

The politics of sewerage

Smell drives the politics of sewerage, at least from the perspective of those who live downwind from treatment plants. More generally, the immense price tag of sewerage construction and maintenance has meant that such projects always depended on public support and political pressure. When these faded, the system was plunged into neglect. The late-twentieth-century politics of sewerage was precipitated by environmental pollution and its adverse impact on health and on property values. In the previous century, the triggers were urban epidemics and the desire for urban and economic progress. The city needed to clean up its nuisances; industry was poised for its second revolution. The bourgeoisie made sewerage a primary item on its political agenda and thus readied a massive reconstruction effort.

When the field of sanitary engineering arose in the nineteenth century and the public handed over its responsibility to the new practitioners, trusting to scientific measures, people grew more distant from the system and became indifferent and reluctant to fund new projects. Sewerage infrastructure slowly fell into disrepair, incapable of handling the staggering increases in volume and generally ignored (with the exception of the depression era, when federally funded projects, such as treatment facilities and plants, were constructed with architectural elegance).

In the late 1960s the American public became uneasy with the growing problems of cities (transportation, air pollution) and more aware of the environmental impact of urban infrastructure; engineers and public works agencies came under increasing fire. Costly new and renovated sewerage constructions, built in compliance with more stringent regulations, required renewed public support. Media reports and political rhetoric during the 1970s convinced people that infrastructure was crumbling and sought more public money to be pumped into these projects. Debate since then has focused primarily on where these facilities should be built and what their size should be, rather than whether they should be built at all.[9]

Advocates of environmental justice targeted the highly political issue of sewage racism in the early 1980s. As development on the edges of cities boomed, sewage plants (like landfills) came nearer to suburban backyards and cities ran out of space on their peripheries. New plants were increasingly built in poor neighborhoods with little political and organizational power. Prior to the 1990s, the damage to property values and quality of life mobilized only well-to-do neighboring communities and environmentalists who feared the known and unknown economic effects of this locally unwanted land use (LULU, in planning jargon). Today, however, a simple siting announcement in even the poorest community can galvanize five hundred people within a day to protest. Essential as sewage plants are, it is becoming more and more difficult to build them unless they are odorless, invisible, and silent. Citizen involvement is necessary to the success of any project. A rarity in the 1980s, a citizen advisory panel with veto powers is now integrated into the planning process early on. Public relations tactics—visual models, computer modeling, newsletters, press releases, tours, brochures, and specially designed visitor centers—play a crucial role as well. The process is often lengthy, with frequent newsletter updates and many meetings. Planners and managers are forced to listen and give assurances; otherwise, long and costly litigation follows.

It has also become necessary to sweeten the deal with aesthetic improvements as well as cultural and recreational compensations. No wonder the headline of the spring 2001 newsletter describing the proposed Bright Water sewage plant in King County, Washington, reads, "Art, open space, recreation and wastewater treatment plants enhance communities."[10] Apparently, the closer the facility is to a residential neighborhood or recreational area (and the greater the sociopolitical power of that neighborhood), the more thoroughly disguised or alternately embellished the facility must be and the longer the process takes.[11] A special mitigation coordinator is appointed to oversee the various measures taken to ensure the plan's approval. To meet fears and concerns not covered by aesthetic and sensory mitigation strategies, some sanitary districts provide a property value guarantee; others offer the community something in return—a park, a community center, or a library. Some communities have proposed a "fair share" law to address the issue of environmental racism and make it hard for rich neighborhoods to reject the siting of a plant in their area (Muschamp 1994, 42). Ironically, the most dramatic public landscapes today owe their existence to the liability involved in waste institutions, and managers of treatment plants are becoming keepers of public space. Some communities are now vying to be selected to host a new sewage plant so that they can enjoy the amenities that come along with it (Leccese 1999, 64).

Another lingering and fierce political debate concerns the highly technological and centralized nature of the sewerage system. Sewer opponents insist that the common approach to sewage is expensive and harmful, degrading more than protecting the environment. They argue that treatment plants are faultily designed and that efforts to recycle byproducts—sludge and water—are futile.[12] They instead advocate small systems that treat black and gray wastewater separately at the source. But the established, capital-and energy-intensive, centralized sewage institution is deeply rooted in the politics and form of the city.

The inherent idiosyncrasies of sewage—the muck, the smell, the language, and the politics—promise that anyone who treads on the noxious physical realm it defines will find a fertile ground.

Terrain of Engineering Novelty

Sewers in premodernity

The sewer is the epitome and embodiment of the modernist project. Surveys of Western sewer history, however, often begin with the Cloaca

Maxima, the great sewer of Rome. The Cloaca Maxima became a symbol for everything civilized and ingenious. Pliny the Elder, a first-century C.E. Roman savant and author of *Natural History,* lumps together in one category the sewers of Rome, the Pyramids, and the Roman Forum. He refers to Rome's sewers as underground navigation channels and rivers leading to the great Tiber River. "Public sewers," he writes, are "a work more stupendous than any; as mountains had to be pierced for their constructions, [so] navigation had to be carried on beneath Rome. . . . For this purpose there are seven rivers made, by artificial channels, to flow beneath the city." (1957, 140).

Constructed in the sixth-century B.C.E. as an open channel lined with stone in an existing streambed known as Forum Brook, the cloaca's primary function was to drain and develop the marshy Velabrum area, which later became the Roman Forum. The initial channel construction, attributed to Tarquinius Priscus, was enclosed in the third century C.E. with a stone barrel vault and drained the stormwater from the Forum; soon, public toilets were connected to the system. When it was finished in the sixth century C.E., the Cloaca Maxima was, according to architectural historian Lewis Mumford, a grand engineering monument of a scale so gigantic that its builders must have seen a long way into the future (1961, 214). It was large enough to admit a wagon laden with hay. Later, another streambed accommodated the Cloaca Circus Maximus, which was constructed on the southern edge of the Forum, south of the Palatine Hill. The two buried cloacas follow the meander of the creek beds down to the Tiber River and have survived the destructive forces of almost two millennia.[13]

For many centuries after its construction, the cloaca remained an anomaly. There were sporadic attempts to bring some order to filth-mired city streets, but no systematic effort was made until the sixteenth century. At that time sewers in many great European cities were often little more than stinking gullies running down the center of the street, which overflowed every time it rained. Built canals and urban waterways became open sewers—which were sometimes covered, as was London's Fleet River in the late sixteenth century. One of London's earlier major sanitary acts from 1388 limited the disposal of fecal matter to designated places outside the city and to specific times, but it was difficult to enforce (Melosi 2000, 40; Kelly 1973, 20). Cesspits continued to flow, fouling all of London's three open rivers as well as the Thames for centuries to follow. A walk on London Bridge became a deliberate act of daring because the river stank so badly.

A delay between recognizing and acting on possible solutions was common elsewhere. Laws enacted in the early 1530s in Paris demanding that each house and every quarter be equipped with a cesspool had little effect. The November 1539 landmark decree, as noted earlier, stated: "We forbid all emptying or tossing out into the streets and squares . . . and we command you to delay and retain any and all stagnant and sullied waters and urines inside the confines of your homes. We enjoin you to then carry these and promptly empty them into the stream and give them chase with a bucketful of clean water to hasten their course" (quoted in Laporte 2000, 4).

Elsewhere, aiming to rid cities of their noxious stench, the artist and innovator Leonardo da Vinci designed a blueprint for urban underground sewers to drain public and private latrines. The sixteenth century was already capped with two critical innovations: the modern water closet, which was first introduced by Sir John Harrington in his 1596 book, *Metamorphosis of Ajax,* and toilet paper, a Chinese invention. "The two of these together moved the cause of civilization ahead centuries," writes Katie Kelly (1973, 20). The visionary ideas of Da Vinci and Harrington were spread during the Renaissance as a new sense of the private and increased sublimation of excrement took hold, though their full implementation had to wait three hundred years.

In eighteenth-century Paris, sanitary reformers referred to the Cloaca Maxima as the most noteworthy achievement of all time and considered it an example of an enlightened state action embodying the spirit of civic patriotism (Reid 1991, 15). The city had a few closed primitive sewers, but these were mostly dysfunctional and easily clogged with mud. An exception was an early privately owned project, the famous 1737 Turgot sewer; it was built as a vault with hewn stone, cleaned by a contractor appointed by the city, washed by water from a reservoir constructed at the sewer's head, and adorned with a garden (Reid 1991, 14).

Formal dumps to hold the contents of cesspools were established outside Paris in the late eighteenth century and persisted for almost a hundred years.[14] Property owners paid a sewer tax to the city, which contracted cleaning firms to dispose of their feces (and animal carcasses). Most of the waste went to Montfaucon, north of Paris, where it was dumped into large settling basins and treated. Liquids would cascade down to lower basins, depositing sludge that was dried and collected into large heaps that sat from one to three years. The pools of fermented fecal matter next to slaughterhouses became a symbol of profanity and immorality and a subject for physical and social reform (Reid 1991, 71–72).

More than any other site, Montfaucon was surrounded by the processes and discourse that represented the realities and aspirations accompanying the transition to modernity. The architect Pierre Giraud was commissioned in the final years of the eighteenth century to come up with a plan for the site. He proposed to transfer the waste and slaughterhouses to enclosed and well-ventilated installations outside the city, where fertilizers, tiles, and other products would be produced from the wastes. The site itself was to be converted and made into a social center in the city. "I would like the dumps, which one does not approach now except with fear and repugnance, to become a meeting place for the curious and for innumerable workers. . . . [W]orkers of all kind will come to breath the fresh air . . . [and] will come to instruct themselves in the Arts which we cultivate in the heart of these establishments" (quoted in Reid 1991, 73). Giraud's plan was not realized until Paris expanded to reach the disposal area and railroad technology afforded an easy access.

In 1844 Doctor Louis Roux concluded, "[Paris] will not be able to merit the title civilized city until Montfaucon no longer exists" (quoted in Reid 1991, 73). In contrast, Professor Jules Garnier claimed that Montfaucon itself should be transformed to a modern industrial facility, an Ammoniapolis, where the waste would be processed and smells would be under control. The unbearable stench finally resulted in the closure of the dump in 1849, and the sewage basins were moved to Bondy. The cesspool barrels were now brought to a clean and ventilated facility in La Villette, not too far away, and shipped via the canal in closed containers to Bondy. Eventually, transformed into Parc de Buttes Chamount, the site became a spectacle of unparalleled beauty and ingenuity (see chapter 3). Cesspools were replaced with sewers in the latter part of the century, and legislation ordered the closure of all cesspools in 1894. All residences were to be connected to the sewer.

In colonial America the first processes and decrees concerning cesspool construction, similar to those in Europe, were set in motion in the late seventeenth century. The cleaning of cesspools was mostly an individual responsibility; their contents were sold to farmers well into the twentieth century. Early sewers, such as those in New York and Boston, date back to the late eighteenth century and were designed for stormwater drainage only. A New York fertilizer company called Lodi Manufacturing was the first of several to produce poudrette in America in 1840 (Melosi 2000, 39). Water closets took their position of dignity only around midcentury, but disposal practices were faulty and recurring epidemics in crowded city centers spawned a decisive scientific and technological response.

Sewerage and modernism

Pipes and plumbing technology afforded the ultimate control and domestication of waste, assigning bodily waste and cleaning to the private domain. The century-long modernist project that began in the mid-nineteenth century entailed the systematic cleansing of the home, yard, and city and the evacuation of all waste and other nuisances quickly and efficiently in a scientifically based network (while soiling the land and water).

The modern Parisian sewer, in particular, captured the imagination of the Western world. The engineer and urban planner Baron Georges-Eugène Haussmann transformed the capital of Emperor Napoleon III both above and below ground. The new urban circulatory system he constructed between 1852 and 1870 became a monument to the Second Empire, free of blocked arteries and foul passages. Haussmann considered the construction of the sewer a "medical surgery" necessary to heal the ailing body of the city. "The underground galleries, organs of the large city, would function like those of human body without revealing themselves to the light of the day. . . . Secretions would take place there mysteriously and would maintain public health without troubling the good order of the city and without spoiling its exterior beauty" (quoted in Reid 1991, 29). He refurbished all Paris streets with sewers—or, more precisely, with multipurpose galleries large enough to suspend other utility pipes from the walls and roof. The wider galleries accommodated sluice cars and boats used to clean them. The sewer put its stamp on Paris's urban form, shaping its axial boulevards. The rivers carrying society's debris below were mirrored in the flowing human grandeur in the avenues above (fig. 5.4).

During the World Exposition of 1867, when the sewer administrators began offering public tours of what the engineers referred to as "un second Paris souterrain," tourists were escorted down the illuminated galleries by sewermen in deluxe versions of the sluice boats to witness an engineering masterpiece of tidiness and order and experience the drama of Hugo's writing (fig. 5.5). (Among the recommended sites for visitors was Parc de Buttes Chamount, the reformed Montfaucon.) The reconstructed Paris sewer thus became a new means to understand the city. "The realization that all the social establishments of the city, be they domestic, commercial, industrial, or cultural, no matter how unrelated they are on the surface, are interconnected underground, excites powerful touristic passions," writes Dean MacCannell in *The Tourist* (1976, 75).

One visitor touring the 1860s sewers of Paris asked, "Couldn't one also decorate the big and small collectors with painted pictures? What a vast space open to the imagination and talent of young artists who dream of big composition" (quoted in Reid 1991, 44). Not until sewers were rescued from their shameful state and made accessible did anyone offer such an idea. Parisian journalists produced fanciful accounts of the tours and enlightened a mass audience about the new world below ground. Artists used paintings and photography to celebrate the technological novelty. Noteworthy was the work of Nadar (Gaspard-Félix Tournachon), who in 1865, after patenting a device to photograph with artificial light, began taking pictures in the sewers. Nadar staged mannequins dressed as sewermen for long exposures in dim light and depicted the sewers' austerity and mystery, enticing citizens and tourists to subsequent visits (Reid 1991, 44). Yet it took another century for public artworks and art exhibitions to reengage and influence the design of sewage facilities.

Across the straits of Dover, English cities had also built sewers under their streets, though with less drama and grandeur. Midcentury urban sanitary theories and practical proposals by health experts Henry Mayhew and Edwin Chadwick to clean the city proved effective and garnered the attention of other Western countries. Indeed, beginning with the invention of the water closet, England served as the hotbed of experimentation in sewage treatment and disposal and exported its innovations across the Atlantic to America.

Looking to Europe on sanitation, mid-nineteenth-century America began catching up and soon took over the lead. The revolution in public sanitation in American cities, precipitated by population growth, established water infrastructure systems, and technological advances in the private realm led to unmatched unified, centrally controlled, buried sewage networks. The early modern, combined sewerage systems (for both human waste and stormwater) were built in the 1850s in New York, Chicago, and Jersey City, New Jersey, though most cities did not begin building sewers until the 1870s.[15] Unlike in Paris, these sewers had little impact on city form. Although Chicago's streets had to be elevated as much as 12 feet to facilitate the flow of the untreated sewage to Lake Michigan, the underground construction did not find an echo in the city plan. When cities discovered that their own water and recreational resources were being severely damaged by raw sewage, they began taking new measures. In 1892 Chicago undertook the largest earth-moving project in the history of municipal public works to preserve the quality of its water supply of Lake Michigan. The regrading of 28 miles of the Sanitary and Ship Canal reversed the flow of the polluted Chicago river and diverted industrial and sanitary wastes via the Des Plaines and Illinois Rivers to the Mississippi River.[16] In 1886 New York built the first of the nation's sewage plants in Coney Island to protect its world-famous beaches. During the first two decades of the new century, thousands of miles of sewer lines and elementary sewage treatment facilities were built. First, the intermittent filtration and later the activated sludge methods were used. Located at the end of the pipe, treatment plants discharged a partly cleansed liquid into nearby waterways and released sludge into drying beds, dumps, and oceans.[17]

The premise behind waste systems was that they should be obscure. While infrastructure for water, transportation, and power—thousands of nineteenth- and early-twentieth-century viaducts, waterworks, reservoirs, and lighthouses—designed jointly by engineers and architects reflected America's pride in its public works and became the envy of the world, sewerage facilities were pushed out of sight. They sank comfortably into oblivion at the same time that Freud was pointing out the dangerous consequences of repressed instincts that lie below the skin and within the body (and by inference below the surface of the city), and artists were expressing their renewed trust of human waste. Late-nineteenth-century realists and naturalists—for instance, Gustave Courbet and Vincent van Gogh—included excrement and dung in their paintings to symbolize the natural cycle of death, decay, fertilization, and rebirth. For them, it was the most elemental form of creation, representing the truth and naïveté of nature and an attachment to the low, the ugly, and the concrete (Chu 1993). Twentieth-century conceptual art used (and abused) the subversive and transgressive powers of sewer and excrement to provoke, break free from taboos, and disrupt class and aesthetic boundaries. Marcel Duchamp, who directly and boldly engaged

the very objects of ordure, made several explicit associations between excrement, money, and art, shocking the art world with his ready-made
urinal titled *Fountain* (1917).

Sewers, however, discussed primarily in professional engineering and
how-to publications, were scarcely apparent in academic accounts, and
the interest of the general population shifted to other, more visible public works.[18] In the 1930s, the New Deal added its notable share of sewerage works, as many cities built sewers; about 65 percent of the country's sewage treatment plants were constructed during that decade.[19]
Artists and architects were employed to design and embellish pump
houses and other structures with decorative details. The Wards Island
Sewage Plant, for example, built in 1937 in conjunction with New York's
World's Fair of 1939 in Flushing Meadows (previously a landfill), boasts
a grand symmetrical layout and art deco–style buildings adorned with
reliefs of the city seal and icons celebrating technology (fig. 5.6). The water tanks are beautifully framed with an elegant operating gallery a third
of a mile long that houses the return sludge pumps. Yet the facility has
remained inaccessible to the public.

After World War II all public works joined the status of the sewer; con-
signed to the pragmatics of the engineer, they were renamed "infrastructure." They were detached from public space, disconnected from
city form, and divorced from public artwork. Until the 1970s Americans,
like members of other Western cultures, suppressed all objects having to
do with bodily functions. Sewage became a technical system involving
scientific predictions of probability and uncertainties regarding urban
growth, rainfall, and water quality. Nevertheless, by the mid-twentieth
century many sewers were stressed beyond their capacity and had become major sources of pollution. In the 1950s every body of water receiving sewage was badly polluted with nutrients and toxins. The public was outraged, and a series of laws to protect water quality came into
effect in the following decades.[20] Freud's warning had materialized; as
their worldview changed, people moved to combat the problems and
shape a new era in sewage history.

The postmodern aesthetics of ordure

The postmodernist project of the 1960s, fed by the politics of the environment and strong public pressure, deconstructed established systems
and offered alternatives. Within two decades, as planning and design of
sewage plants were reexamined, tight engineering standards were forced

to make space for art and design considerations. Sewage began to take on an architectural presence. The new imperative was driven by a critical cultural discourse taking place in the art world. Artists engaging the subjects of scatology—the study of excrement and bodily ejaculations—made a previously private subject public and focused attention on basic instincts, things "common to all" (Weisberg 1993, 18).

Following the tradition of dada, the situationists, and pop art, scatology in the early 1960s and increasingly in the 1980s and 1990s became a distinct field in contemporary art. Its most renowned representative is Piero Manzoni, who canned and offered his own excrement for sale in *Merde d'Artista*. The price per can was based on the current price of the same weight of gold (fig. 5.7). Influenced by the theories of Freud, Marx, and Pierre Leroux, Manzoni drew on the relationship between excrement and gold and excrement and resistance. Another artist, George Maciunas, created an inventory and recorded his defecation patterns, pinpointing the normality of shit. The artist Dieter Roth pressed rabbit feces into a quaint chocolate Easter bunny mold, contrasting bodily functions with commercialism and reproduction. These tactics were aimed both to attract attention and to scandalize audiences. Beyond a desire to shock, claims the art critic Gabriel Weisberg, these artists sought to achieve "artistic freedom that shattered boundaries created by traditional standards of taste and decorum" (1993, 18). It is only when all aesthetic aspects can be studied without prejudices that the veil surrounding excrement and sewage is removed and the subjects can be normalized.

Art exhibitions through the 1990s continued to develop these ideas. In 1994 high art was first housed in a sewage facility. The exhibition titled *Cloaca Maxima* curated by Hans-Ulrich Obrist at Zurich's sewage plant exposed and reframed the undignified subject of human civilization and shit, considering the themes of sewage, scatology, and the toilet in a political context (Babias 1994). Among the artworks on display were chamber pots, historical and contemporary toilets and bathroom paraphernalia, wastewater pipes, sewer signs, and other objects related to defecation.[21] During the exhibition, tours of the sewage treatment plant took visitors to chambers filled with unbearable odors.

Meanwhile, ecological artists and designers, concerned with environmental pollution, began taking a more practical approach to sewage systems. They advocated decentralizing the sewer system, personalizing commitment, and encouraging less-consumptive habits. A pioneering art proposal (whose assumptions have become normal practice twenty years later) was *Municipal Water Gardens* (1969) (fig. 5.8). Patricia Johanson submitted this alternative to the conventional sewage system to *House and Garden* magazine, together with a number of other conceptual garden proposals. Combining art with functional infrastructure in an attempt to address environmental problems, *Municipal Water Gardens* proposed to turn a sewage treatment plant into a public water park, in which six lakes, shaped like a flower's petals, recycle wastewater within a closed system. Wastewater first enters a lake of effluent; as it flows to a second lake it passes through a sand and gravel filter, and as the water turns progressively cleaner it becomes available for human use. The third lake allows boating; the fourth lake supports aquatic communities and fishing; and the fifth lake is safe enough for swimming. Finally, the water passes into a reservoir to be used as drinking water. The design strategy considers both problem and solution as part of the same cycle, and it makes us see the potential good in sewage (Johanson 1993, 3).

The artist who more than anyone else has made wastewater a centerpiece of his artwork is Lewis (Buster) Simpson. In the social activist tradition of Joseph Beuys, Simpson incisively questions the very assumptions built into urban sanitation and sewage systems and takes on public sensibilities. In 1978 he positioned himself smack in the muck and made sewage outfall visible while measuring effluent flowing into rivers in major cities and testing their toxin level. To draw connections between the two ends of the digestive system—food and sewage—he placed concrete casts of picnic plates at sewage outfalls in the Niagara River and then displayed them with their stains. In 1984 he initiated *When the Tide Is Out the Table Is Set* (an old Salish Indian expression indicating the abundance of food at low tide), in which porcelain dinner plates, made at the Kohler toilet factory, were placed for a year in major effluent points (fig. 5.9). When fired in a kiln, the wastewater contaminants glazed the plates in vibrant colors and beautiful patterns, which became indicators of toxins; the more toxins in the water, the prettier the plates became (Matilsky 1992).

Composting Commode (1991) epitomized Simpson's ongoing urge to break down the conventional, centralized urban networks of sewage and drainage and to make the management of waste more personal and local. In 1987 he designed and built a composting toilet for downtown Seattle as part of a public art program. The portable, plastic, ventilated, and handicapped-accessible commode was to be placed on a busy downtown street over an empty tree pit; when it filled up, it was to be moved to the next pit. The trees later planted would have enjoyed a supreme soil medium, and the facility would have addressed pedestrians' need for public toilets. Simpson spent three years trying to obtain permits from public health and public works departments to operate an apparatus formally described as a "potential health hazard" and informally perceived as an offense to genteel sensibilities, but no luck (Clifton 1993, 25). A tree marks the first site of the commode on the street, which was served for only two weeks. Eventually, after a few years of disuse, the commode was placed in a Seattle school garden where students learn about waste management and recycling while tending the plants. The project also initiated a public discussion about the need for public restrooms in the city.

These artworks and exhibitions have stimulated an unpopular public discourse, exposed deep-seated assumptions, and disrupted aesthetic conventions. They proved that the modernist project to separate the city from the resources that sustain it is harmful, and they pointed out practical approaches toward a renewed and closer connection between the public and public space with sewage facilities.

The new architecture and art of sewage plants

Opening up a previously unvisited subject, European sewage plants of the 1980s began to display conspicuously the industrial character and sculptural attributes of their massive structures, highlighting the visual prominence of the large anaerobic tanks.[22] For example, the Munich Gut Marienhof sewage plant (1984–88), designed by Kurt Ackermann and Partners, intentionally contrasted the facility with the surrounding landscape and accented the plain geometry of the utilitarian buildings. The plant became an attraction, and the silver pyramidal digestors were its symbol (Knapp 1998, 41) (fig. 5.10). In the revitalized Emscher region of Germany, six new sewage plants were the subject of a design competition in the early 1990s. The winning design for the Bottrop treatment plant shows off its stark silver cylindrical digestors that rise like alien spaceships from the flat landscape. In the sewage plant in Göteborg, Sweden, an impressive entryway and a complex of pools and gardens complement the massive architectural volumes and welcome visitors to the site (Suneson 1999).

Other daring designs of sewage facilities from around the world in the 1990s suggest that the economic, technical, and aesthetic considerations of the facilities have merged, taking on conspicuous, industrial elegance of form and material.[23] But as Europeans began treating their facilities as public works monuments and visible urban landmarks, Americans engaged in sophisticated (or not so sophisticated) camouflage tactics. Many artists and architects who joined multimillion-dollar public sewage plant projects entered an increasingly complex and paradoxical process, which often turned them into tamed public service providers.

Even the best architects and engineers failed, ultimately making exceptional efforts to keep the offensive structures secret. For example, in 1967 the architect Philip Johnson hoped to turn the New York North River sewage plant into a city monument, with a giant water sculpture on the roof. Instead, it was buried twenty years later under a roof that carries a community recreation complex (see below). In the late 1990s, San Diego's newest facility—the North City Water Reclamation Plant, bounded by freeways and an army base—was designed to create a business park atmosphere while preserving the character of the site's natural surrounding. The city is proud that people often mistake the facility for a community college.

Percent-for-art programs managed by city arts commissions became a driving force behind the integration of public works and public artworks in the 1980s and 1990s and garnered some success. These organizations bestowed on artists the role of mediators between the municipality and the public. At times budget constraints and ineffective leadership merely produced a publicly acceptable image, providing little public access and only gesturing toward art with minor ornamental elements or a sculpture. One example is *The Hydrofiter* (1992), a symbolic and interpretive 30-foot-high kinetic sculpture that functions as a fountain and a weather vane, which was built by Evan Lewis in the 91st Avenue Wastewater Treatment Plant in Phoenix, Arizona. At other times, projects are accessible to the public and offer new spatial and architectural manipulations, resulting in an elegant and humane environment for employees and visitors.

In some facilities, sewage pumps located at various critical points of the system are transformed through art into local neighborhood markers.[24] Sewer culverts, too, were reconceived as artworks and turned into promenades. Both Patricia Johanson and Buster Simpson had an opportunity to design sewer culvert boxes. Johanson was invited to "save" the construction of the Sunnydale sewer pipe line and pump station along

Candlestick Cove's shoreline in San Francisco.[25] Her 1987 proposal, *Endangered Garden,* buries the sewer line and realigns it, using its cover, 30 feet wide and one-third of a mile long, as a baywalk. It is shaped and colored after the endangered San Francisco garter snake and is surrounded by a series of small gardens, tide pools, and a butterfly meadow. Finished in 1994, the sewer line is a utility sculpture and an ecosystem meant to restore and nurture habitats of endangered wildlife (Johanson 1992, 22–24; 1991). Unfortunately, despite the visible and evenly spaced manhole covers along the promenade, the design makes no reference to the sewer below.

Simpson's unrealized public art proposal for the Danube Canal in the city of Vienna also engaged a sanitary box culvert, then under construction along the river edge. *A Dialogue along the Danube Canal* (1994) was intended to make the invisible visible, to bring sewage to public notice, and to inspire a discussion about attitudes toward sanitary utilities while creating a meaningful, singular experience along the culvert and canal corridor (Simpson 1994). Simpson made the sewer line an integral part of the experience. The plumbing is exposed and the concrete box becomes the surface of a promenade and a stage for multiple fountains and sequential hydrokinetic events. Restroom facilities, called Culvert Commodes, were to stand atop the sanitary box culvert to acknowledge directly its function and provide a much-needed service.[26]

Retired sewage plants offer yet another design opportunity that operating plants cannot afford, an unrestricted access. Following cleanup of toxic soil substances, their massive built structures can be adapted for new, unforeseen uses and unimagined forms while vividly responding to the facility's past use. Luckily, the high demolition costs are often responsible for channeling decisionmakers into creative solutions. Three examples begin to sketch some possibilities.

The conversion of a retired sewage plant in Bellingham, Washington, into a maritime interpretation center and fish hatchery, uses the actual sewage pools and the inherent aesthetics of the sewage plant's architecture in the new facility. The old structures are fully converted into pools and streams, and the site operates as an urban park and learning center where visitors can fish steelhead and cutthroat trout and watch salmon spawn (Masterson 1981).

Bellingham serves a precedent for the current design by Carol R. Johnson Associates of Cambridge, Massachusetts, for the defunct Sturgeon

City Sewage Plant and its adjacent landfill near Jacksonville, North Carolina. This 27-acre area will become a museum and a park in which all existing structures retain their structural identity and are reused. Building on previous groundwork and plans by locals and the North Carolina State University School of Design, the design team is transforming the industrial building and treatment tanks into interpretive exhibits on marine life and aquariums to raise fish and the old sludge-drying beds into plot gardens "while educating about use, misuse, and healing of this coastal landscape" (Sorvig 2002, 44).

The landscape architecture firm of Wenk Associates of Denver, Colorado, has recently converted the Northside Sewage Treatment Plant, 50 acres of drying flats and holding tanks into a city park. The design retained the plant's remnants mostly as reminders and support structures. The plan makes available a broader range of permanent park elements that would otherwise have been economically unfeasible (Leccese 2001). Partially buried structures and drying beds have become multipurpose playing fields, exposed structures are reused as shelters and viewing platforms, the old concrete channels are planted with wetland species and recreational facilities, and excavated fill material is made into striking earthworks.

The projects of the 1980s and 1990s began to open the giant sewage machine to public gaze. They put to the test engineering standards and assumptions, safety codes, and public tolerance of human waste. They proved that sewage facilities are as much a design problem as an engineering problem. But they have yet to exhaust the design possibilities of these institutions. Public access into most facilities remains limited, and little progress has been made in transforming unfavorable public sentiments or rethinking the relationship of the sewage system to the city, public space, and adjacent community. A newly instituted process of political consensus about prospective projects have, at times, posed hurdles and hindered creativity. Architecture, landscape architecture, and art become indispensable means to appease local opposition, though they have also been used to make these debased structures merely palatable. Artists and designers are caught up in the ethics of disguise, conciliatory tactics, and minuscule embellishments. Landscape architects are called on to help soften the scale of the structures, "integrate things," and "minimize the visual impact."

The character that we have come to know and dislike—pure functionalism, industrial monumentality of forms and lines, magnitude of task and process, unexpected juxtaposition of identity and locale, and surreal eeriness at night—has been neither fully accepted nor aesthetically exhausted.

The sewage plant's multifaceted nature—as a locus of essential utility, intense feelings, and contradictory meanings—makes it a rich subject for cultural and design exploration. The questions remain: What possibilities exist for designers and artists to transform this unique entity into a distinctive form of infrastructural public works? What new programmatic combinations of architecture, landscape, and engineering could normalize it, bring people closer to the sewage system, and make it safer environmentally, viable economically, and agreeable socially?

Sewage Prospects: Projects and Critique

Although many new plants proclaim their environmental friendliness and civic responsibility, only a few reflect and enhance the interlacing conditions of the urban and natural systems. Even fewer embrace their bold technological character as a powerful resource to build on, to strengthen their formal coherence, and to draw attention to their aesthetic qualities.

Several design and art projects of sewage plants, divided into five groups, have been selected for extended discussion. The first group illustrates the camouflage approach that discourages public access. West Point in Seattle and Oceanside in San Francisco are hidden under concrete roofs and behind planted berms and terraces. They deny their connections to the city and adjacent public spaces. The second group of facilities takes the utilitarian approach and offers a public space but still suffers from the confining aesthetics of shame. The designers of North River in New York and Donald Tillman in Los Angeles made considerable efforts to invest the sites with quality architecture, public space, and recreational opportunities, but they did little to reveal the essence of the sewage plant. The third group attempts to open the plant's grounds to public viewing and engender a positive, more direct interface with the community. Newtown Creek in Brooklyn, New York; Columbia Boulevard in Portland, Oregon; and Marine Park Water Reclamation Facility in Vancouver, Washington, offer the public broad access, a public garden/park, and a visitor or community center and observatory to view the plant. Plants in the fourth group employ the educative and celebrative strategies. The 23rd Avenue facility in Phoenix and Point Loma in San Diego invite (or plan to invite) the public to a choreographed tour through their grounds and offer a unique experience of the operation. The fifth and final group—California's Arcata Marsh and Wildlife Sanctuary; the Renton, Washington, Waterworks Garden, and the 91st Avenue sewage plant in Phoenix—includes sewage plants that integrate wetlands as part of the cleansing system, merging wastewater machines and wastewater gardens, turning the working plant into a sustainable, integrative work and leisure ground.

Existing in disguise

The prevailing agenda for sewage plants of the last two decades has been disguise. When in the city, they are partly buried underground, supporting usable space on a roof; when in the country, they are completely covered with earthforms and plants, preferably creating a large park.

Oceanside Wastewater Pollution Control Facility. Oceanside, the newest sewage facility in San Francisco, combines impressive architecture, an enclosed, outdoor public space, and elegant details, but it turns its back on the city and the ocean. Its spatial and architectural features designed by CH2M Hill engineers and Simon Martin-Vegue Winkelstein Moris architects are a visual treat for glossy magazines and those few visitors and employees who venture through the entry tunnel into the hidden canyon. The project design team shared the view that on completion, they "didn't want to be aware that there was a sewage treatment plant there—didn't want to see it, hear it or smell it" (quoted in Dawson 1990, 60). The realization that public infrastructure can no longer be devoid of art and human quality resulted in award-winning architecture that is worth seeing. However, the plant's beauty is hidden in a concrete canyon surrounded by sound-deadening retaining walls and planted berms; it is revealed only to organized public tours. All the operating tanks and machines, except for the four egg-shaped sludge digestors and the administration building, are buried under the berms and the roof, leaving space above for future expansion of the adjacent city zoo.

Surprise and awe await only those who enter the plant through the long entry tunnel. Visitors are welcomed by a central open space and parking area surrounded by 45-foot-high retaining walls; the main building has elevated terraces and walks. Walkways take people to the top of the berms, which afford views of the entire concrete canyon (fig. 5.11). Artwork, required of public buildings by the state's percent-for-art program, was conceived by the design team and appears mostly on the retaining walls and in the public area. Conventional, precast retaining wall surfaces are scored to allude to the shape of hoppers within the building, pierced to create porthole-shaped gardens, and scoured to suggest erosion caused by rushing water. A fountain fed by reclaimed water emerges from a porthole-shaped niche in the wall next to a public area laid with various paving surfaces, furnished with benches, and dotted with small trees. Hardy native and nonnative plants, selected by the landscape architects Royston, Hanamoto Alley, and Abey to withstand the ocean winds and salt spray, create a grayish blanket over and around the plant. The landscape architects were especially involved with the siting and, of course, the greening and screening of the facility.

Oceanside has been featured in numerous architectural magazines, and the city takes a great pride in its state-of-the-art investment (Schmertz 1994; Leccese 1995). However, hidden from view the new facility missed an opportunity to create a new public space that would have opened to the highway, beach, and ocean and combine two interesting urban institutions—a sewage plant and a city zoo—while keeping the neighborhood side green.

West Point Wastewater Sewage Treatment Plant. Seattle's coastline has been the home of the Westpoint sewage plant since 1966. The plant's recent expansion, also designed by CH2M Hill engineers in collaboration with the landscape architecture and urban design firm Danadjieva and Koenig, takes the camouflage approach to new heights. The design cloaks a large part of the plant in a restored native landscape. Structures were designed not to intrude or tower over the surrounding forest of Discovery Park. The plant is cut into the hillside, and the landscape advances onto the concrete roof with the assistance of tiers of retaining walls. An undulating, planted "China Wall," as the designer calls it, buffers the plant from the coastline and the walking trails in the forest above (fig. 5.12). The braided and textured bleached-white concrete walls recall the bluffs and the flow of water, creating a soft, sinuous topographic layout. At one point, the wall is lifted and becomes a bridge that frames the plant's entrance.

Lead landscape architect Angela Danadjieva, who obviously worked under the pressure of community groups seeking to eliminate the plant's negative effects, promised park users that they will never see the $500 million sewage plant (Powell 1992, 39). They do see, however, the $10 million worth of native plants that are planted in a large restored park around the sewage plant. The surrounding park, which opened in 1995, features trails, rolling earth berms, a beach, forested hills, and a constructed wetland that receives and treats the runoff from the plant and hillside areas. A short time after the planting was complete, all of the plant's structures had vanished within a thicket of alders, maples, wild roses, and beach grasses. The real drama—a startling clash of machine and planting—is hidden inside the plant. The thirteen giant circular secondary clarifiers are set against overgrown planting beds and host occasional waterfowl. The facility's north section, where aeration basins are buried below ground, has become a thick wilderness punctuated by vents and access structures.

The award-winning project has been heralded as a great success. "The design turned what might have been an eyesore into a shoreline retreat . . . and it is even possible to forget what is happening a stone's throw away," reads a feature typical of many found in architectural magazines that covered the project (Gantenbein 1995, 59). Defending her camouflage approach, Danadjieva claims that she merely healed a disturbance, a temporary scar in the landscape.

Oceanside and West Point missed an opportunity to use public works to enrich public space, to create a dialogue between seemingly contradictory functions, and to rethink the connection between the way that our landscapes work and look. Plants that are disguised or that divert attention from their actual operation pay only lip service to being environment and citizen-friendly.

Estranged public amenity

Sewage plants built near residences must address the community and emphasize the interface between the plant and the neighborhood. Not surprisingly, the plant's fence is often the focus an art project. This highly visible element, which defines the boundary between inside and outside, does not interfere with the operation of the plant and, for some, is potentially a screening device. For the expansion of the Coney Island Water Pollution Control Plant in Sheepshead Bay, Brooklyn, the New York artist Ned Smyth devised a sculptural divide titled *Wave, Wall, and Green* (1991) that is made of a chain-link fence and covered with native plants. It depicts a series of waves that rise and fall and eventually breaks onto the sidewalk. Two other plants responded more generously to the community and added some public amenities—a boardwalk, a recreational park, or a garden. Unfortunately, the amenities still remain se-cluded and divorced from the plant.

North River Sewage Plant. The North River Plant, located on the Hud-son River in Manhattan, bears a 28-acre landscaped roof. In 1962 New York chose the derelict rail yard in impoverished, space-starved Harlem between West 137th and 145th Streets over more affluent locations to dump its waste. Five years later Mayor John Lindsay hired Phillip John-son to design the plant with special consideration to its visual impact from the New Jersey side of the river, the George Washington Bridge, and the adjacent elevated Henry Hudson Parkway. Johnson proposed to en-close the plant in plain geometric volumes and thereby create a mod-ernist sculpture, a new monument to the city. The roof was to become a water display that celebrated the water-cleansing process below. Hun-dreds of aeration nozzles forming a continuous shimmering haze of wa-ter droplets, four powered jet fountains, and reflecting pools were to be viewed by motorists on the abutting parkway and from high-rise apart-ments (fig. 5.13). An alternative design shows the parkway lowered and covered with a roof filled with recreational fields that mediate between the neighborhood and the same water display.

The idea of a sewage plant turning into a monument and a water sculpture, an unparalleled precedent at the time, was instantly con-demned by the community as a vulgar irony, but the notion of an ele-vated, connecting park lasted through subsequent proposals. In 1969 the architects Gruzen and Partners were asked to utilize the entire rooftop for a park that would accommodate a long list of recreational facilities. Gruzen's proposal avoided any reference to the facility, concentrating in-stead on connecting the rooftop to the neighborhood and to Riverside Park along the Hudson River in order to preserve the integrity of the shoreline park. In the following two decades, while the foundations of the plant were built over the water (on hundreds of caissons pinned into the bedrock), the design went through a couple more iterations.

In 1986, after twenty-four years of bureaucratic wrangling, struggles for funding, and intense community negotiations, the plant opened buried under a roof and behind a concrete veil of a pierced arched facade that stretches for ten blocks and faces the river. The facade is reminiscent of other arched bridges and of a latter-day Roman aqueduct. Theodore Long of TAMs, a consultant who was responsible for the outer appear-ance of the plant, was charged with the task of creating a pleasing image with "no flag waving" (Iovine 1988, 119). The structure was completed in 1991 with Richard Dattner's $129 million roof park, the only state park in New York City (Dattner 1995). With its lavish sportsplex, swimming pools, gym, outdoor rink, running track, promenades, pavilions, amphitheater, ball courts, and cultural buildings, it immediately became popular (fig. 5.14). The architecture critic Herbert Muschamp wrote that Harlem residents had the right to insist that the plant be out of sight but blames the architects for lacking in creativity and producing an uninspiring project (Muschamp 1994). The tame cover of the public space avoids any visible and physical connection with the plant below. The only clue to its presence is provided by fifteen tall smokestacks, which protrude through the roof and exhaust plant gases. Views of the facility are available only from the carousel pavilion area, from the two freestanding stairways that connect the roof to Riverside Park below, and from the amphitheater area on the lower platform. Only those descending to the public amphitheater, which is nicely embraced by the plant structure, can gain direct views into some of aeration tanks and the loading dock. The design has been primarily criticized for breaking the continuity of the park along the river, and some have proposed bridging the gap via a boardwalk-promenade.

If designed with public access, the facility could have become New York's grand sewage museum, offering a twenty-first-century experience equivalent to the nineteenth-century grand tour of Paris sewers.

Donald C. Tillman Wastewater Reclamation Plant. Another winner of architectural awards, the administration building at the Donald Tillman plant in the San Fernando Valley of Los Angeles is surrounded by reflecting pools, waterfalls, and fountains and houses an indoor visitor center. It adjoins an extremely popular Japanese garden that is fed with reclaimed water from the plant. Apparently living in harmony with its working neighbor, the garden keeps secret its source of life.

The plant was constructed in the early 1980s and became fully operational in 1985. The administration building acts like a partition between the two dramatically opposed parts of the facility—the wastewater plant and a surreal, Japanese garden designed by Dr. Koichi Kwana (fig. 5.15). This peculiar combination was the idea of Donald Tillman, the city engineer, who conceived the plant and became deeply engaged with Japanese garden design after taking a course on the subject. The concept, which the community endorsed, served to pacify local opposition. One of the glossy brochures featuring the garden states that "The authentic Japanese garden provides an opportunity for visitors to experience the beauty and tranquility of another culture while learning more about wastewater treatment and reuse" (City of Los Angeles Public Works 1995, 6). Visitors to the garden can easily bypass the bordering sewage plant, the garden's reason for existence and its water supplier. Despite the sweet smell that emanates from the lake and the facility's announced purpose of "demonstrat[ing] a positive use of reclaimed water," Mr. Green, the garden manager, proudly declares that very few visitors even realize there is a sewage plant behind the wall (interview with author, Los Angeles, July 1995).

Considering the popularity in America of parks with exotic and multiple themes, the combination of a Japanese garden and sewage plant should not surprise anyone. The escapist, foreign-themed garden seems to have won people's hearts, managing not only to sweeten the sour pill but also to create a perfect illusion in which the presence of the sewage plant simply disappears. The appeal of the garden is evident. It has been the setting for more than 200 feature films, commercials, and television movies and for more than 350 weddings and other ceremonies. People stroll along the curvy ponds, rock arrangements, carefully pruned planting, and Japanese lanterns on flagstone paths and bridges on their way to the tea house. Only those who take the time to climb the concrete bridge or the catwalk that connects the building and the garden arrive at an observation point that overlooks the sewage treatment pools on the other side of the fence and across the road.

Garden and sewage could have conspicuously merged to expose the duality of the site, generating a much-needed dialogue between two public urban entities. The tantalizing fact that the plant is linked and supplies water to the river and to newly constructed lakes in the distant regional Balboa and Wildlife Parks is mentioned only in brochures found at the information center. If allowed access, people could have experienced firsthand the connections between the effluent, the river, and the garden. This fascinating and complex human-made system remains instead invisible and intangible.

Urban gateway into the environment

A few sewage treatment plants, some recently built and some being planned, invite the public to explore their grounds and provide a window into the workings of the city and its water resources. An early proposal by the artist Christy Rupp, working on New York's Coney Island facility, focused on the interface between the plant and Shellbank Creek intended to make this edge accessible to people via a promenade starting at the street and to develop a wetland. Along a 400-foot-long boardwalk, Rupp proposed a fence titled *Tidal Filter Fence* (1991), whose wavy look simulates the lapping of the tide and whose screen and detail alludes to the screening and filtering taking place in the wastewater purification process. The accompanying boardwalk was planned to describe the function of a proposed tidal wetland and to bring attention to the plant. Rupp's proposal was favorably received by the New York City Department of Environmental Protection, but the project is on hold. Because plants commonly bound the water's edge, a promenade or boardwalk is an element found in several art and design proposals. Although it remains outside the plant, it more actively invites the public to experience the relationship of plant and water.

Newtown Creek Water Pollution Control Facility. Brooklyn's Newtown Creek facility is undergoing a major expansion and renovation through 2010; this includes improvements in architecture and public space that will invite the community to pay attention to the plant and bring the facility closer to the urban waterfront. The 1967 sewage plant abutting the working-class neighborhood of Greenpoint is located among derelict industrial and waste facilities at the edge of Newtown Creek, a murky tidal inlet of the East River.

The entire facility is being rebuilt, and two new buildings are being constructed, disinfection and support facilities designed by the architecture firm of Pholshek and Partners and engineering offices of Greely and Hunsen, Hazen and Sawyer, P.C., and Malcolm Pirnie, Inc. These have already won the New York Arts Commission award for their elegant and bold composition and vivid use of materials and colors. Brightly colored glazed brick will identify the function of each building and provide a contrast to the giant steel and concrete structures (Nobel 1999, F13). The disinfection building will be located at the northern point of the plant by the water's edge, over the contact tanks that serve as the last cleansing point before effluent is discharged. A transparent elevated walkway with a bright yellow skeletal structure will connect the support building to the other side of the plan. The lighting designer Herve Descottes of L'observatoire International will create a theatrical night scene that floods the entire plant in a soft bluish veil. The blue light, signifying cool, calm water, contrasts the orange color of the city and will provide a dramatic view from Manhattan. According to the design concept, "the infrastructure will not be serving our daily convenience without being recognized" (L'observatoire International 1999) (fig. 5.16).

The percent-for-art budget is divided between two landscape art projects. Each addresses the interface between the plant and the neighborhood and is located at the edge of the site.[27] The environmental artist George Trakas designed a quarter-mile promenade park that starts at Paidge Avenue and follows the northern edge of the plant toward the waterfront. The walk connects the neighborhood with the disinfection building and provides much-needed access to the creek. Plantings, seating areas, outlooks, bank treatment, and interpretive elements of the industrial and cultural history of the area adorn the walk.

Acconci Studio, led by the artist Vito Acconci, is working on the south edge along Greenpoint Avenue and the entry to the visitor center/administration building. Acconci's main idea, "The city reaches into the plant; the plant receives the city, and the people in the city," shapes the design (Acconci Studio 1997). His 1997 preliminary proposal was far-reaching in scope, implausible and ludicrous for the time. Acconci envisioned the sidewalk and fence as long arms stretching and invading the plant like tree branches, inviting passersby toward the plant. The fence deforms into cylinder, tubes that function as tunnels and trellises for climbers. The walkway forms ramps over the plant, which hover above aeration tanks; "people become surveyors onto the workings of the plant" and can stop for picnics at grassy areas. At night it glows with light like a comet bursting over the plant (1997).

The initial proposal was tamed, scaled down, and restricted to the periphery of the plant. Acconci Studio was awarded the 1999 New York Arts Commission award for its revised proposal, *Edge of the Plant/Edge of the Neighborhood/Mix of Edges* (1998–2000), which drew particular praise for its respect for the surrounding and for the way it has brought the neighborhood into the plant (Nobel 1999, F1). In the new proposal the fence continues to probe and negotiate with the plant grounds, giving space back to the neighborhood. The fence and its accompanying concrete wall act as a place maker, creating small parks and pools of water as they coils around the property line. A glass wall penetrates the visitor center in the administration building and creates, both inside and outside, pockets of public space that dissolve the border (fig. 5.17). As Vito Acconci explains, "The wall/fence, therefore, becomes a tool to interfere, invert, and disrupt the status quo, as well as an invitation to enter, a path leading into the plant" (interview with author, New York, January 19, 2001).

The disparity between the first and second proposals reveals the limits on access that still exist and brings to light the difference between the theoretical and the attainable. Undoubtedly, the plant will become more transparent and emerge as a community resource as designs respond directly to the critical issues of access and edge conditions between neighborhood, plant, and water.

Columbia Boulevard Sewage Plant. The Columbia Boulevard plant, situated between the Portsmouth neighborhood and the Columbia Slough in Portland, Oregon, daringly invites people to enter its headworks building and gain a panoramic view of the facility. The 1993 plan for the plant upgrade by CH2M Hill engineers included an extensive landscape design component by the landscape architecture firm Murase Associates. The design aims to create a front door to the facility, connect it with its adjacent neighborhood, and open up its peripheries to the public via trails. The result is an easily accessible facility that is visible from the road. A new gate and entry sign, a tree-lined road and parking lot, and a garden welcome the visitor to the headworks building and the interpretive center it houses (designed by Miller Hull Partnership and Sera Architects).

Despite the purpose of opening the facility to the public, plantings along the periphery were clearly used to screen "less pleasing elements," "create a naturalistically planted perimeter," and "fit the plant into its surroundings" (City of Portland 1993, 8–1, 8–3). The highlight of the design is a water garden that is pushed against the headworks structure, reflecting its steel and glass facade (fig. 5.18). A series of shallow ponds and a marsh treat and aerate water that is diverted from the secondary clarifiers and later used for local irrigation. A footbridge crosses over the working marsh and leads to the interpretive display at the headworks building. People are then directed up to the third floor and via the screen bars shafts to a dramatic panoramic overlook of the plant. The landscape masterplan ties the facility to the riparian strip along the Columbia Slough and the 40-mile-long path on the north bank, as well as to a new park on the west.

Marine Park Water Reclamation Facility. Situated along the Columbia River, the town of Vancouver, Washington, went as far as building a community center at their sewage plant ground, making the place one of the most popular local gathering sites (fig. 5.19). In the early 1990s, when the facility needed to expand and upgrade, the community voted to keep the waterfront plant in place but turn it into an accessible public resource. The Marine Park Water Reclamation Facility was rebuilt (by CH2M Hill) and its functions were enclosed in a series of small, red brick buildings with blue pitched roofs in a typical neotraditional suburban style. Housed in a large, pseudo-Shingle-style building, the Water Resource Education Center contains an exhibit hall, art gallery, classroom, multimedia theater, computer room, and water science laboratory to monitor and study water quality, as well as the office of the city's Cultural Services Division, which seeks to integrate the arts and sciences. The center overlooks the Marine Park, which once again gives the public access to the 48-acre restored wetlands and riparian zone and includes a riverfront trail and a wetland outlook. A growing number of volunteers and a crew of "aquaguides" offer tours and organize community projects, special events, lectures, and workshops. The facility received the 1997 American Public Works Award and has become the centerpiece of the community, giving it a new image (APWA 1997).[28]

In an age of shrinking land resources, the luxury of placing "undesirable" facilities at a distance from residential, commercial, and recreational areas no longer exists. The sewage landscape, positioned at the waterfront boundary between the city and its water supplies, provides an entry to the ecosystems and communities through which it passes and becomes a threshold, a critical gateway into the environment.

A public art tour

23rd Avenue Wastewater Treatment Plant. The desert city of Phoenix, which has long recognized that its water (and, more recently, waste) resources are precious, allows residents a more thorough experience of the sewage plant's operation. Just four blocks north of Phoenix's renowned garbage facility, in a peripheral industrial zone of the city, is the 23rd Avenue Wastewater Treatment Plant, whose public art scheme opened up its grounds to the community. In 1991, the Phoenix Public Arts Commission selected two artists, Laurie Lundquist and Jaune-Quick-To-See Smith, to develop a proposal that utilized the percent-for-art funds from the plant's capital improvement program. Their rejected proposal featured a 45-minute public tour that would wind through the facility's structures past interpretive signs and installations. The tour was direct and humorous, filled with visual and other sensory content. Geared for schoolchildren, it substituted quick, strong, visual gestures for lengthy verbal explanations of the water reclamation process. The original control building, a small 1931 Spanish colonial–style building, was preserved for use as an information center and a departure point for the tour (Cameron 1994, 21). According to Lundquist the project was voted down because "there was far too much contact with the wastewater staff" and it was far too light-hearted and possibly threatening to the committee. "They wanted something more scientific, straightforward, and high-tech" (telephone conversation with author, September 16, 1998).

In 1997 the project was revived, and the new commission was given to the San Francisco artist Su-Chen Hung. Hung's intervention, which builds on the early concept in a more symbolic and abstract form, was accepted and completed in 2002. *Water Spells* choreographs a journey that starts and ends at the proposed visitor center and follows the six stages of the water-cleansing process. " 'Wastewater' is an ironic term, implying discarded and useless water eliminated from the life cycle. Nothing could be further from truth," writes the artist. "My goal here is to raise visitors' awareness of the process of 'water-treatment' to a new level—to celebrate rather than to ignore a process long considered ignoble and disagreeable at best" (1997, 1).

Hung marks the cyclical nature of water through seven installations and narrative signs that emphasize the recovery of water and materials from wastewater and make the treatment and the usefulness of the materials recovered in each phase vividly transparent to the visitors. Topiary of trailing lantana forming the word "water" in Latin, Hohokan (an indigenous Indian language), Chinese, and English is planted at the entry area and irrigated with the clean water that leaves the plant. A recycled drip water screen cools off the porch of the visitor center. A sculpture spelling the word "water," made of clear resin into which solids recovered from the plant's screens are embedded, marks the first station of the tour. At the aeration basins a 5-foot-tall installation demonstrates the oxygen-based microorganism process that breaks solids in the water (fig. 5.20). A totem-like transparent cylinder divided into six distinct sections, presenting water collected from each step in the process, marks the place where the final chlorine is added (and will be reproduced on a smaller scale as an inexpensive souvenir). People returning to the visitor center at the end of the tour can enjoy a restaurant and sit beside a fountain in the Spanish-style courtyard.

Regrettably, the installations rely heavily on narratives inscribed on signs and are inconsistent in style; some are cartoonish, others highly refined. The massive size of the machines and water tanks sometimes overpowers and diminishes the installations. Moreover, the tour avoids any reference to the sludge and the dewatering facility integral to the workings of the plant.

Point Loma Wastewater Treatment Plant. Artworks are also integrated into the 1996 comprehensive blueprint for the expansion and improvement of the Point Loma sewage plant in San Diego. The plant, located outside the city, joins a navy base and the Cabrillo National Monument on a coastal cliff overlooking the ocean. The project, a joint effort of the City of San Diego Metropolitan Wastewater Department and the Commission for Arts and Culture, brought together a team led by the artist Mathieu Gregoire—two architects, a landscape architect, a civil engineer, an ecologist, a poet, and seven other artists and designers specializing in color, sound, imagery, light, photography, and graphics. The design emphasizes the workings of the plant and at the same time resolves the site's conflicting uses: sewage, military, and recreational. The design concept visually promotes the mechanistic and industrial character of the plant by making the new and renovated structures more transparent and by weaving the landscape and the machine so that the two "begin to form a texture rather than an inadvertent dislocation" (Gregoire 1996, 2).

The plan also serves as a guide for future expansion and upgrades. The landscape design recommendations emphasize the facility's connections to the surrounding landscape in terms of habitat restoration and plantings. They also propose a pedestrian network, including entryways, paths, and common spaces for employees and visitors. Unlike other designers on the team (and contrary to the concept of maximum transparency), the landscape architects used screening techniques to "correct and soften" some views that are considered unpleasant (Gregoire 1996, 6). The architects focused on developing a simple, flexible module for new buildings and a consistent scheme to tie new and old structures together through form and color; they stressed transparency and lightness for maximum visibility, safety, ease of maintenance, and aesthetics. The artists and designers further sought to enrich and amplify the experience of the place through elegant and surprising interventions in structures, spaces, and views. Classical and modern images of nature and technology are designed for central outdoor walkways and indoor terrazzo floors and mats. A rhythmic poem etched onto handrails, sidewalks and walls, and glass is designed to run in and out of buildings, walkways, and pipe galleries. Humorous haikus, meant to relieve the stress of employees, have been proposed for certain underground walls. Despite the design intent, the aesthetics of some of the proposals do not draw on the for-

mal prescriptions of the site's machinery and may result in mere em-
bellishment. Conversely, the neon strip lighting in horizontal fixtures
and attached to tanks, the upper rims of ramps, and retaining walls
promises to accent and unite the plant structures in a nighttime scene.

Unfortunately, the few projects that are already in place—the sludge
pump facility, two digestors, and an administration-lab building—bear
little resemblance to the guidelines. The architects, Metcalf and Eddy, designed enclosed and opaque structures with ornate emblems. Nearby, a
section of 12-foot-diameter outfall pipe is displayed. Meanwhile, landscape architects and ecologists labored to restore the damaged 6-acre
steep hillside above the plant and added a huge berm. Some justify the
berm as defining a "gateway" to the plant; others see it as obstructing
views of the plant from the highly popular south point of the peninsula.
The improvement plan promised to shape an industrial site of bold character and great public appeal that would draw visitors to experience the
plant, but its implementation remains uncertain.

A vision for a sewage plant could seek its justification within the bold
geometry of buildings, tanks, and pools, which closely resembles the
great geometric formal gardens of the seventeenth-century French master André Le Nôtre. Designs could draw on the dynamic spectacle of function, the existing processes of filtration, sedimentation and aeration, separation and dilution, reuse and recycling. The ongoing drama of the
mechanical water ballet provided by hydraulic pumps and fountains and
the surreal lighting quality could simply be emphasized. The pressure of
fountains could be linked to the volume of water being processed, with
colors used to signify different kinds of pollution. The extreme sensory
environment could be put to use, stimulating olfactory sensation via a series of enclosed "smell rooms" and fragrance gardens. Planting and gardens, the antithesis of waste, could be employed not to cover but to uncover what goes on in the place. Rather than setting the machine in the
garden, gardens could nestle in the machine and prompt a cultural dialogue between flowers and dung, the living plant and the mechanical
plant, cleanliness and dirt. Plants could open their laboratories and control rooms that rely on high-tech visualization media to the public, and
they could transmit information from various other sites about water volumes and quality. Designs that employ irony, inversion, and wit could
offer relief, giving a comic twist to these inherently serious and thorny
matters. Eventually, design and art could achieve a position from which
to influence how we should treat our private waste and public water.

This last agenda is made more explicit in the final group of sewage facilities.

Wastewater gardens

Replacing purely mechanical sewage plants with an integrated wetland-mechanical system has recently become a compelling option for wastewater treatment. Machine and garden hybrids that together clean raw sewage unite people's love for living plants and gardens with the practical benefit of treatment plants. Aquatic ecosystems—marshes, swamps, and bogs, which we lump together under the term "wetlands"—now participate in secondary and tertiary treatments. The juxtaposition of the two working landscapes has created a new aesthetics and public landscape. These wet landscapes permit free access to enthusiastic crowds of bird watchers, hikers, and nature aficionados. Much like the late-nineteenth-century Parisians who flocked to the productive sewage farms in Clichy and Gennevilliers, today's urbanites are attracted to sewage gardens that are both culturally and ecologically productive. At times, these plants assume new roles and act as laboratories and learning centers, as monitoring systems and feedback mechanisms.

This combination of nature and mechanics, which blurs the lines separating work from leisure and technology from culture, presents new opportunities for our supermodern cities. Towns are choosing the hybrid approach because of its low cost as well as its ecological and recreational advantages.

Scientists and environmental engineers began experimenting with biological treatment systems in the late 1960s. Bill Wolverton discovered the phenomenal capacity of aquatic systems to filter and remove water pollutants as he studied the Florida marshes and later on built the first wetland of water hyacinth for the NASA space lab in Bay St. Louis, Mississippi.[29] Such wetlands, often called the "natural kidney" or "marsh ecology," act as an effluent polishing system, mainly for tertiary treatment; they have been successfully implemented in numerous sites in the United States and around the world. Toxic substances, heavy metals, nutrients, and bacteria in the water leaving a facility are broken down by plants, microorganisms, fish, and sunlight, in a constructed wetland or a series of small lagoons planted with marsh plants (cattails, reeds, rushes) or vascular aquatic plants (duckweed, water hyacinths) through which the water slowly flows to a main waterway. These create rich wetland habitats. A modified system called a "rock marsh," in which the roots of plants in a rock medium cleanse the wastewater that flows below the surface, was devised for both cold and warm climates. These systems can fit in a small backyard and treat heavy-duty septic tank effluent; larger installations can serve towns of more than 25,000 people. Many macrobial rock plant filters exist in southern states and contain flowering plants such as calla and canna lilies, bulrush, cattail, and elephant ear. Visually, these sewage treatment plants are indistinguishable from a flower garden and can even support profitable flower farms.

Arcata Marsh and Wildlife Sanctuary. The first town to adopt the aquatic plant system was the small town of Arcata, California. Built in 1986, the wastewater landscape complex—containing 50 acres of oxidation ponds and aquaculture (fish hatcheries where trout, salmon, and other fish are raised and then released into Humboldt Bay) and 40 acres of constructed wetlands, known as the Arcata Marsh and Wildlife Sanctuary—surrounds the town's sewage plant, which resembles an island of built structures amid a wide-open water landscape.

The cost- and energy-effective project was needed to comply with stringent water-quality standards and to protect the marine life of the bay. It eventually produced the most important community asset: a park where people engage in recreational activities, study ecology, and track birds. The park is the focus of various events and "marsh and treatment works" tours. Each marsh bears the name of an individual who has contributed to the project's success. (One, now called "no name marsh," still awaits its name.) An interpretive center run by the Friends of the Arcata Marsh (FOAM) was built overlooking the entire area and provides the usual community activities and exhibits, particularly with ecological or educational themes. The town and the marsh have become synonymous.

Waterworks Garden. The Seattle artist Lorna Jordan, who found out about the Arcata marsh, created Waterworks Garden just outside the East Division Wastewater Reclamation Plant in Renton, Washington, to purify the stormwater from the 40-acre plant area.[30] It demonstrates the power of plants to filter water, using the formal language of gardens that cannot be mistaken for that of nature.

Following six years of negotiations with the community and the plant engineers, who viewed art merely as decoration, Jordan completed (with the aid of the Seattle-based landscape architecture firm Jones and Jones) the 8-acre functional earthworks/waterworks in 1996. The project was funded by the King County arts program and by the plant's stormwater treatment expansion budget. Jordan's early proposals to use various forms of water to create temporal landscape experiences, such as a cloud machine emitting fog on the hilltop or a fire-breathing artwork fed by methane gas from the plant, were rejected. Jordan's breakthrough occurred when she linked the idea of a working marsh with the plant's stormwater requirement, which initially was designed as a cluster of three rectilinear fenced ponds. Instead, she conceived of a water garden shaped like a flower, symbolizing growth and sustenance.

The garden is reminiscent of Patricia Johanson's 1969 *Municipal Water Gardens,* in both concept and form. "The idea," claims Jordan, "could best relate to the engineers' work and treat people as participants rather than spectators" (interview with author, Seattle, April 24, 2001). The slurry water pumped from the plant grounds begins its journey at an elevated pond framed with ten columnar rocks, pours to eleven terraces with leaf-shaped ponds where contaminants and sediments settle, trickles through a grotto, cascades to three other ponds, and finally flows to a wetland before it drains into Springbrook Creek. Instead of a plain, scientific model of wetland restoration with didactic labels, it employs spaces and references drawn from traditional gardens and myths (Leccese 1997). The grotto, laid with mosaics and shaped like a seedpod, is embellished with concrete stalactites, small pools, and a fountain. Planned tours of the plant include the garden, which is maintained by the Metro, the agency that handles Seattle's wastewater treatment. The garden is much loved and heavily used by the community. As with other gardens of this sort, its ultimate success is signified by the desire of many couples to marry or take wedding pictures there (Canty 1998). The power of the garden resides in its working relationship with, and visual and mental contrast to the conventional plant (fig. 5.21). Unfortunately, a tall wall separating the plant and the garden prevents most people from experiencing the union of the two facilities.[31]

91st Avenue sewage plant. Pressed to comply with new regulations, more facilities in the United States are putting wetlands to use. The giant 91st Avenue sewage plant in Phoenix, Arizona, is about to construct 800 acres of wetlands to upgrade its operation. The wetlands will provide partial secondary and full tertiary treatment before releasing some of the water to the dry riverbed of the Salt River.[32] The existing 12-acre demonstration project known as Tres Rios (Three Rivers), built in 1994–95, has already proved its success not only in purifying wastewater but also in attracting waterfowl, amphibians, insects, and humans. Tres Rios draws large crowds on weekends. It offers a range of facilities, such as bird-viewing blinds, horse-hitching posts, shelters, and handicapped-accessible walks, for its recreationists, those on educational tours, and overnight campers. The expanded wastewater facility both creates an oasis in the desert and a much-needed wildlife and human refuge and demonstrates wise use of water, promising to enliven the dry Salt River belt.

In the 1990s many other places with sensitive aquifers and coastal and marine resources around the world utilized the marsh ecology for tertiary treatment.[33] Evidently, these wastewater infrastructure projects have been responsible for the most recent massive "constructed nature" projects. But must these landscapes imitate nature as we have always known it? Do we have an opportunity (and good reason) to constitute a new kind of nature and aesthetics that express the new conditions? The landscape designs evolving from scientific-engineered solutions insist on being read as nature as it is commonly understood; thus, in the absence of verbal interpretation their content and process remain obscure. The Arcata marsh looks like a natural marsh, and the sewage pipes and interconnectedness of the parts that make up the total design solution are hidden from view. Consequently, their meanings are hidden as well. In contrast, the wastewater cleansing gardens of Lorna Jordan conceive new forms that unite nature and human technology.

The new artificial wetland is a form of sewage treatment more complex than any of our present mechanical technologies. These facilities reflect the recent discourse and new agenda of supermodernism, which will mature in the years to come and endow infrastructure with a new role. Sewage infrastructure has always been utilitarian; now it has become self-sustaining as well. Sustainability and practicability are merging into one. Utility systems and services are being decentralized, designed with overt aesthetics true to their function and reassigned to the everyday public landscape. Environmental engineers, designers, and artists now lead the way in taking on a new task.

Living Machines and Other Exploratory Scenarios

Time magazine's special issue "Beyond 2000" (November 8, 1999) predicted that in the future all sewage will be piped into high-efficiency marshes enclosed in flexible structures supported by light skeletal membranes, where selected plants, fish, snails, and microbes will purify the wastewater before it enters streams and reservoirs. No longer will inadequate treatment of wastewater promote algae bloom that threatens other aquatic life (Alexander 1999, 115). Even in dense urban settings with limited open space, biological treatment systems are feasible. Bill Wolverton envisions a multistory structure, a skyscraper of plants, where effluent is purified as it trickles down to the tap. Pioneering solar aquatic technologies have paved the way for various urban applications that combine wastewater treatment with urban agriculture and aquaculture. John Todd and his living technologies research team at Ocean Arks International have developed an early prototype, called a bioshelter. It consists of cleansing tanks in an enclosed greenhouse-type structure and produces many valuable by-products, including aquatic plant feed that can be fed to animals and fish and ornamental plants, biogas, and electricity that can be harvested and sold.[34] Todd envisions the solar aquatic system as a community-run enterprise managed by members in greenhouses attached to the sunny side of buildings, on balconies and roofs, and in empty lots (Todd and Todd 1994, 48).[35] It requires 75 square feet per person.

An improved version of the bioshelter, the Living Machine, is a compact mobile unit, which treats wastewater to secondary treatment standards. It is made up of self-contained ecosystems, usually a tank divided into a number of cells or a number of tanks filled with rock media. The substrates provide a home for plants and billions of water-purifying bacteria, microorganisms, zooplankton, snails, clams, crabs, and fish that live within and cleanse the effluent as it moves slowly through the cycle. There are more than twenty Living Machines in eight countries treating from 4,000 to 1,000,000 gallons a day (fig. 5.22). At Oberlin College's Center for Environmental Studies in Oberlin, Ohio, the Living Machine is housed in a solarium adjacent to the college auditorium and serves as a working classroom for sustainability; it is integral to the design of the center and accompanied by an outdoor constructed wetland.[36] A Living Machine garden space with a cultured butterfly community enclosed by a screen structure is similarly integral to the design of the visitor center of the National Audubon Society's Corkscrew Swamp Sanctuary in Naples, Florida. Visitors walk through it on their way to the toilets.

Another early prototype of living technologies designed to clean sewage in dense urban centers is called the sewage solar wall. This small-scale treatment facility is placed directly in the street and is connected to adjacent apartment buildings. The sewage walls run along the city block and separate the street from the sidewalk. The walls, capped with glass, channel sewage through a series of terraces filled with plants and bacteria that progressively filter and purify the waste. The resulting effluent could be used on local gardens, and the plants harvested and composted. Whether in tanks, in greenhouses, or along sidewalks, biological treatment systems offer a productive alternative urban landscape.

The restoration of polluted wastewater lagoons and open canals (or large-scale water bodies as lakes and reservoirs) generates other types of landscapes—floating gardens anchored in a raft complex. The Restorer, patented by Ocean Arks International, is an ecological filter that pumps and lifts polluted water into planted microbial beds, which break down contaminants. Solar panels or windmills can be used to power the onboard compressors. The floating islands break free from shore and migrate around the water surface with the changing winds. As water and sediment pass through a series of reactors, excess nutrients are digested and the presence of organisms beneficial to a healthy ecosystem is enhanced.

Finally, artist and engineer Viet Ngo has had great success using the duckweed plant (Lemna, in Latin). He refined a bioengineering technique to recover polluted water areas relying exclusively on this plant (fig. 5.23).[37] The duckweed absorbs harmful chemicals from the water, multiplies rapidly, and can be harvested and composted or fed to animals. In an earlier project Ngo devised a 60-acre functional wetland with a four-mile long serpentine earthwork planted with duckweed in Devil's Lake, North Dakota. The earthwork and wastewater ponds take on an artificial form inspired by indigenous Indian burial mounds and suggest the shape of an intestine. The giant garden is open to public education and houses a laboratory, a modest viewing dock, and paths along the dikes, but it did not acquire the public grand, leisure stature Ngo had hoped for.[38] Ngo refers to his work as water gardens for waste, places that inform, soothe, and delight the senses and nurture birds and waterfowl.[39]

Increased use of composting toilets and graywater systems in the private realm and more local bioswales that treat stormwater would significantly reduce sewage volumes and make mechanical sewage plants participants in a more complex natural-cultural system rather than the sole players in the game. Gardens, greenhouses, flower farms, micro-fish farms, and tree farms would join sewage systems in creating ecologically sustainable and productive urban landscapes. These ideas for a decentralized, local, efficient, and inexpensive wastewater treatment will transform the city and bridge the gaps between people, infrastructure, and the environment.

Victor Hugo and other social visionaries targeted the sewer as a locus of sociopolitical critique and used it as a tool for recovery, for a better tomorrow. Environmental engineers sought out new, more effective and economical treatment technologies. Artists freed excrement and sewage facilities from their negative, moralistic associations and engaged the fetid subject in order to question boundaries and shake old habits and taboos. They taught us that like other kinds of refuse, bodily waste is a creative realm that, when opened up, allows freedom. They used it as a basis for inquiring into the relationships between body and politics, private and public, city and environment.

Sewage and excrement expose the multivalency—both strength and weakness—of our lives and our cities. Sewage facilities can be sites that are unashamedly beautiful: places that express the fullness of life, simultaneously conveying what is gray and debased, as well as what is precious and refined. Sewage systems will become contributors to the purification of the city and symbolic receptacles for what the purified space and deodorized discourse left out.

Challenges for Thought and Action

On Monday, August 27, 2001, Secretary of the Interior Gale Norton declared the garbage dump of Fresno, California, which was the first sanitary landfill in the nation, to be a National Historic Landmark. A few hours later, the decision was rescinded—that the dump previously had been designated a Superfund site, a hazard to groundwater and people, was somehow overlooked in the landmark-naming process. Major newspapers around the country covered the incident, seemingly entertained by the flap. A *Los Angeles Daily News* headline read "Something Rotten on Bush's First List of U.S. Historic Sites." As the dump's honor was dumped, Fresno's local paper proclaimed: "garbage in, garbage out for Fresno."

The lively discussion that has followed reveals intense feelings about both decisions. Designating a cultural scorn as a historical landmark seems pointless to many, including environmentalists who fear that the title would honor a symbol of wastefulness and pollution and thus glorify the culture of excess. David Orr, the director of Field Programs of Living Rivers, laments, "And most of these structures have become or will become toxic time bombs, for future generations to deal with. This is a sad story of the landfill culture in which we live, and that is the story that I hope will come back to haunt Gale Norton."[1] Historians, in contrast, see the designation as neutral, merely marking significant cultural structures and places. After all, they point out, not all history and not all existing landmarks are wonderful and pristine. What matters is the accompanying interpretation, the need to make all voices and sides heard. Other avid proponents, the Society of Industrial Archaeology and the Public Works Historical Society strongly favor the landmark status, arguing that it would finally put an end to the out-of-sight, out-of-mind approach. "Perhaps we should simply designate all our superfund sites as landmarks," suggests the historian James Williams.

Although the controversy made visible the paradoxes and ironies entangled in the history of waste and sanitary engineering, the choice of a landfill as a landmark suggests that Americans have, to some extent, come to terms with this formidable landscape and overcome fears, since we now consider sanitary utilities to be significant public structures. While questions concerning the form and experience of a historically designated landfill, or possibly a sewage structure, still remain open, environmental designers and artists should be ready for this new opportunity.

Another paradox is inherent in the current proposals to create a memorial to the victims of the World Trade Center tragedy on top of Fresh Kills landfill, at the site where the remains of the towers mixed with human remains were brought to their final resting place. For example, the design of landscape architecture and urban design firm Field Operations proposed an earthwork mound surrounded by grassland, while that of landscape architecture office RIOS Associates proposed a forest that victims' families would plant for their loved ones. The sacralization of a previously debased site and the mix of cultural cast-offs and revered human remains raise yet other questions regarding the relationship between cemeteries and dumps and between material culture and humans. Beyond dumps, recycling centers, and sewage plants, there are waste landscapes that pose greater and more severe challenges and questions.

We are now facing a new frontier; the daunting toxic landscapes that pose serious dangers and challenge our science and technology. They harbor a far more deceptive waste—the kind of waste that flows with rivers, shed tears into underground waters, clouds the air, and blows across the arid western desert as gas or as dust weighing microns, millionths of a gram. This destructive energy is the transmutation of our immature technologies, which have outgrown our knowledge and capabilities. The science that mined radioactive elements, splicing and dicing them into new, more dangerous elements, has created for itself an epic task, and scientists are in a constant sprint to catch up with our toxic waste. The wastes of massively scaled industries during the military-industrial buildup of the twentieth century become in the twenty-first century the cancers of our earth, and possibly beyond.

The monster we created takes many forms: the black and rust steel ruins of the Ruhr Valley where the German war machine was built, the toxic slag heaps of mines in Montana, the vast million-gallon underground tanks storing radioactive waste on the Hanford Reservation in eastern Washington, the bombed-out deserts of Nevada, or the hidden, spoiled ponds of New England. These destructive energies carve physical scars on the earth and genetic scars on human genomes. The monster lurches in urban and rural fringes and faraway lands. We may get a whiff of it as we travel along the highway on our way to a cleaner, gentler existence. Looking down from above to see black-green bodies of water seeping along, creating exotic shapes and colors, we may wonder about the monster. But what will we do when it catches up to us, to our water pipes and our developing suburban landscapes? What places are we making to contain this monster? When we reach the former margins of our communities, will we skip over these wasted sites, build higher fences to ensure isolation; or will we solve the problem, clean the soil, water, and air, and provide a secure world for future generations?

Radioactive wastes are currently in limbo, waiting for the completion of permanent storage sites.[2] The temporary storage of these wastes has now become a matter of grave concern. Power plants from Minnesota to Florida are piling up waste with no plan for its secure storage. Our current solutions to the liquid storage of this radioactive excrement have taken the tradition of shunting to new heights and depths.

The Waste Isolation Pilot Plant (WIPP)—the first storage site established in the United States, near Carlsbad, New Mexico—buries long-term radioactive waste in an immense salt deposit below arid desert scrublands. But during the coring process, the site was found to be unsuitable for high-level radioactive waste. As the hunt for a high-level waste disposal site continued, only one answer was found: Yucca Mountain, Nevada.[3] For its deadliest garbage dump, the Department of Energy in the early 1990s employed not only scientists and engineers but also anthropologists, linguists, archaeologists, paleoastronomers, and designers to invent markers that would signify and communicate the danger of the buried content at the site to civilizations ten millennia into the future. As a way of bypassing linguistic and communication problems, the architect Michael Brill and his team developed a series of landscapes made of archetypal forms that signify universal threat, including a field of thorns, a barren, scarred land, and a moat. Ultimately, the ideas from several teams were used, and a berm and a series of 25-foot-high granite monuments inscribed in seven languages were built (Kastner 1999). Critics ridicule the underlying concept, pointing at our absolute inability to predict interpretations and motives so far into the future (Erikson 1994). The true ethical question, they say, involves the very practice of producing this kind of waste in the first place and probes human nature and limitations.

Military sites abound. The testing of atomic and nuclear weapons in deserts and on remote islands leaves fields of artifacts, shrapnel, and ideological fingerprints. The bomb fields of Nevada at the Yucca Mountain rim are wastelands created in the instant of combustion. The blast of the bomb and the spray of steel flying through the air and landing and morphing the land create a silent, disfigured landscape, a new metamorphic extrusion of culture. Resembling a devastated movie set or a desolate moonscape, it stands as a poignant reminder of the destructive energies of our military might. This is a waste product, a ruin that some believe worthy of preservation and display. Photographers such as Richard Misrach, Peter Goin, and David Hanson have brought images of these restricted areas to light (Goin 1991; Misrach 1990; Hanson et al. 1997). Their lenses provide the critical distance needed in imaging and imagining the results of an obsessive pursuit of military power, while the average tourist is kept from venturing about the fields and away from the revelations they contain.

In a proposal for the creation of a national park at the site of the U.S. Navy bombing range Bravo 20, Richard Misrach designed a tourist destination with a strong political statement packaged in military and destruction themes. It includes a "devastation drive," a "boardwalk of the bombs," "primitive camping," as well as "an earth-covered concrete visitor's center resembling an ammunition bunker, and a museum devoted to the history of military abuse in peacetime" (Sutro 1993, 83). His collaborator, the landscape architect Pamela Burton, concentrated in her proposal on the site's "context and philosophical implications," avoiding commercialization and disneyfication (83). Proponents argue that preserving bomb ranges would reveal and record the conscience and memory of the atomic age.

Standing on top of a giant landfill, the waste theorist in Don DeLillo's *Underworld* tells a landfill engineer colleague of his forecast and vision for future garbage disposal and landfills:

> The more toxic the waste, the greater the effort and expense a tourist will be willing to tolerate in order to visit the site. Only I don't think you ought to be isolating these sites. Isolate the most toxic waste, okay. This makes it grander, more ominous and magical. But basic household waste ought to be placed in the cities that produced it. Bring garbage into the open. Let people see it and respect it. Don't hide your waste facilities. Make an architecture of waste. Design gorgeous buildings to recycle waste and invite people to collect their own garbage and bring it with them to the press rams and conveyors. Get to know your garbage. And the hot stuff, the chemical waste, the nuclear waste, this becomes a re-

mote landscape of nostalgia. Bus tours and postcards, I guarantee it. (1997, 286).

Radioactive waste sites can become hot tourist destinations and ecological preserves that guard flora and fauna from the destructive impact of more conventional human development. The 310-square mile Savannah River Site in South Carolina, is a decommissioned Defense Department facility that produced tritium and plutonium for warheads. Replete with buried radioactive waste tanks and contaminated lakes and rivers, it was designated by the Atomic Energy Commission as "America's first National Environmental Research Park" in 1972. It became a nature reserve and the home of a research center specializing in radioecology and genotoxicity—the study of radioactive migration, absorption, and effect on ecosystems and animal genes. While not significantly affecting the reproductive patterns and life-span of species, writes John Beardsley, "It would be one of the great ironies of the cold war if mothballed factories for weapons of mass destruction turn out to be among the safest heavens we can offer the nonhuman species with whom we share our landscape. Call it the postnuclear wilderness" (1998, 143).

Seeking solutions to the problem of plutonium disposal, the *Bulletin of the Atomic Scientists* in May 2001 called on artists, architects, and visionary thinkers to enter a competition to design a "Plutonium Memorial." The memorial is to provide safe storage for the world's plutonium for eternity, a place "where tourists can visit" that is "beautiful and grand and awe inspiring."[4] The winning entry by Michael Simonian of San Francisco titled "24110" (the number of years it takes for half of the atoms in the plutonium mass to decay), located the memorial south of the White House in Washington, D.C., (as opposed to Nevada) under a partly lifted elliptical lawn "carpet." The siting turns upside down two scared ideologies: the out-of-sight (and in the poor's backyard) and the great American lawn cover-up. Humorous and witty, it states that sweeping the issue under the carpet is no longer an option.

The difficulties and paradoxes raised by waste have forced us to rethink ethics and aesthetics; they have also stimulated environmental scientists, engineers, artists, and designers to free the spirit from the limitation of professional conventions and the chains of social taboos. Their efforts promise to make a place for the creation of new landscapes that are healthy and strong in form, function, space, and meaning.

Notes

Introduction

1. The book briefly mentions but avoids extensive discussion of incinerators. Specialized toxic waste dumps, resulting from industrial or military activity, are also not part of the book.

2. The works of the cultural geographers J. B. Jackson, D. W. Meinig, Pierce Lewis, John Stilgoe, and others have legitimized the academic study of the ordinary landscape, paving the way for a meaningful scrutiny of previously overlooked landscapes.

CHAPTER 1. Contemplating Waste

1. London had an impressive water supply system that brought water into the city back in the Middle Ages. Private companies began connecting wealthy residences to water pipes only in the early eighteenth century (Melosi 2000, 40). The first American waterworks facility was built in Philadelphia in 1801.

2. Though synonymous, these terms have slightly different connotations. "Garbage" commonly refers to solid material that includes wet or organic discards (mainly household waste but also all municipal solid waste). "Trash" refers to any dry discards. "Refuse" and "rubbish" are terms that include all rejected, unwanted matter. "Junk" is any article that is worn, of poor quality, or discarded. "Litter" is trash or garbage that intrudes conspicuously in the landscape.

3. "Man's twenty-sixth excretion is himself," claims Enzenberger (1972, 8).

4. Only in unreal living places are we able to achieve the total lack of garbage. The fictional worlds like Walt Disney World approach such a condition. What we cannot see is a giant underground pneumatic disposal system that whisks fifty tons of garbage a day from fifteen garbage collection points scattered over the fantasyland to a remote central incinerator (Kelly 1973, 110).

5. The concept of entropy was proposed by the German physicist Rudolf Clausius in 1865. In Greek, *entrope* means change.

6. See Kendrick's discussion of the evolution and authorship of old scriptures and classic texts, in which she explains how revisions, notes, and textual commentary in the margins supplemented and corroborated the text, thus making it possible to replace the script less frequently. The gradually accumulating marginal notations eventually were incorporated into the body of the page, thereby changing the text itself (1992, 837).

7. Most scholars agree that marginal individuals or social groups belong partly to two differing societies or cultures but are not fully integrated into either; instead they are excluded from or exist outside the mainstream of a society, group, or school of thought.

8. On a quite different note, ecologists have always referred to the physical margin, also called the ecotone, most favorably. From an ecological perspective, margins are desirable conditions. The margin, an area between two different ecological niches, is a zone of particularly rich habitat and high productivity.

9. America in particular has mastered the creation of wastelands. In their book *Derelict Landscapes* (1992) John Jakle and David Wilson probe the pervasive phenomena of dereliction in America, pinpointing the cultural-economic considerations that led to decaying urban central neighborhoods and defunct industrial sites. They link the processes to fundamental American values, such as individualism, obsessive mobility, search for better lands, and the objectification and commoditization of land.

10. The original publication "Of Other Spaces" appeared in *Diacritics*, Spring 1986.

CHAPTER 2. **Private Landscapes of Waste**

1. Hygeia was the goddess of health in Greek mythology. In the seventeenth century "hygeia" stood for "a system of sanitation or medical practice." The term "hygiene" was used more widely in the eighteenth century to imply the knowledge and principles for preserving health. A century later it acquired a greater scientific aura (*OED*, 1989).

2. For an in-depth discussion of the sink, see Lahiji and Friedman (1997); for the tub, see Giedion (1948); for the toilet, Horan (1996) and Wright (1960); for public bathrooms, Kira (1976).

3. The discussion here centers on the middle-class urban and suburban landscapes, although some discussion of the wealthy is necessary. The residences and yards of the well-to-do have always served as a model for imitation by the middle class. Waste practices and places for lower-class and rural residences diverge greatly from those examined here.

4. Excavations of farmstead sites from the seventeenth and eighteenth centuries in the Chesapeake region of Virginia reveal trash pits close to various outbuildings in which food was stored and processed (Linebaugh 1994, 11).

5. Philadelphia was an exception, as Benjamin Franklin established the first systematic garbage pickup service in 1792 (Kelly 1973).

6. This theory blaming human sources for epidemics was held by the "contagionists"; it was well established by the turn of the nineteenth century,

7. The modern water closet is attributed to Sir John Harington, the god-

son of Queen Elizabeth I. In 1596 he wrote the *Metamorphosis of Ajax,* outlining the essential components of the toilet—a stool pot, a water tank fed by a cistern, and a flushing mechanism with a brass valve. The book was an entertaining satire on the habits, both repulsive and fanciful though sanitarily ineffective, of humans. Alexander Cummings patented an improved valve closet in 1775, and Joseph Bramah perfected the flushing mechanisms in 1778. Bramah's model ruled the market for 120 years; final improvements to it were made only in the 1880s.

8. The pan closet was a hinged metal pan, which kept a few inches of water at the bottom of an upper bowl. When a handle was pulled, the pan tipped the water and content into the trap below.

9. The practice of covering and replacing filled-up privies with new ones was now prohibited in most large cities, and private scavengers under city contract or city employees engaged in some periodic cleaning. However, services were still inefficient and irregular (Schultz 1989, 167).

10. The White House installed a bathroom in 1851 (Bushman and Bushman 1988, 1225).

11. For examples, see *Carpentry and Building* 1 (February): 27 and 1 (July): 124 (New York: David Williams, 1879).

12. I refer to the 1880 Clean Water Act and the 1890 Smoke Pollution Control Act enacted by Congress.

13. *Plumber and Sanitary Engineer* 1, no. 1 (1879).

14. See *Carpentry and Building* 7 (April): 75–77 (New York: David Williams, 1890).

15. The quote, by the New York LHPA is originally from *Memorial to Abram S. Hewitt on the Subject of Street Cleaning* (published by the association in New York, 1887, 4–5).

16. In 1880 less than half of urban streets were paved. Existing pavements ranged from gravel or macadam to cobblestone and granite block. In choosing street paving types, the emphasis was on sanitary qualities, drainage, and gravity flow of subsurface utilities as well as the ease of travel. Mobility had become important to a new leisure class, as seen in their trendy pastime of bicycle riding. By 1924, municipalities had paved almost all urban streets (McShane 1979, 4).

17. Emmons Clark, secretary of New York's Board of Health, declared in 1891 that privies and cesspools had been eradicated from the city (Stone 1979, 294). Although privies were eliminated, septic tanks were still used in yards for homes not connected to the sewer system.

18. Over half of the entries eliminated the alley as an unnecessary and unsightly adjunct that wasted much valuable area. Fourteen instead incorporated common back areas, such as gardens, parks, and playgrounds (Yeomans 1916).

19. The Waste Reclamation Service, created in 1917 as a section of the War Industries Board and transferred in 1918 to the Department of Commerce, initiated many recycling programs in selected communities (Hoy and Robinson 1979, 4).

20. The ultimate space and cost saving was embodied in Buckminster Fuller's 1938 prefabricated bathroom and kitchen units. These, however, were not favored by the middle class, who looked for greater flexibility.

21. The 1928 Radburn Plan was the first to recognize the garage as an adjunct to the dwelling.

22. Technologies for garbage collection and disposal greatly improved only after World War I. Most vehicles used until the war were horse-drawn (Melosi 1981).

23. In the 1950s, some cities estimated that 25 percent of all garbage was ground by the in-sink disposal (Melosi 1981, 207).

24. The usual 6 to 8 gallons per flush can be reduced to 3.5 gallons in a low-flush toilet, to 1.6 gallons in the ultra-low-flush toilet (ULT), and to less than 1 gallon in the micro toilet (Kourik 1990).

25. It was based on the 1930s Clivus Multrum, a brand-name designed by the Swedish engineer Richard Lindstrom that has become synonymous with today's composting toilet. It is a one-chamber sloped-bottom composter containing two baffles and air ducts (Del Porto and Steinfeld 1999, 3).

26. The Advanced Green Builder Demonstration Center of the CMPBS is a laboratory for the experimentation of sustainable building technologies that the architect Pliny Fisk III has pioneered. Everything from foundations, to columns, to walls, to the roof, and the water, energy, and sewer systems are designed using local, low-tech, recycled materials manufactured with sustainable techniques. (Lerner 1997).

27. The demonstration building of the Toronto Healthy House (1991) contains a closed system in which wastewater is collected from sinks and showers and recycled into washing machines, dishwashers, and toilets. It is then treated on-site and returned back to the cycle. Accessed from http://healthyhousesystem.com (12/1/01).

28. California was the first state to allow subsurface graywater irrigation in 1993 (Del Porto and Steinfeld 1999, 181).

29. This method is called the Honey Wagon/BARC method (Fisk 1993).

30. Every year, the average American uses 6,263 gallons of valuable drinking water to flush 1,300 pounds of excrement.

CHAPTER 3. Dumps

1. In 1865 *Harper's* magazine first used the term "dump" to describe waste piles by mines in the Rocky Mountains (Breen 1992a, 13).

2. "Heap" comes from Old English *héah,* meaning "high"; it implies a collection of things thrown one on top of another, a pile. A "tip" is a deposit, usually with a pointed end. "Tail" refers to the back, last, lower, or inferior part of something; the term is used in conjunction with industrial waste piles, especially those shaped in an elongated narrow shape.

3. Congress passed the 1965 Solid Waste Disposal Act, which authorized federal research, training programs, and grant support for projects that provide alternative to burning of solid waste. Amended in 1970 and retitled the Resource Recovery Act, it prohibited, among other things, the practice of open dumps and it enforced the sanitary landfill method. It also emphasized recycling, resource recovery, and waste to energy practices and encouraged research and the development of new solid waste technology.

4. The well-publicized wanderings of the *Mobro,* which had to return to Long Island after being turned away from six other states and three other nations, brought unmatched attention to the garbage problem.

5. A landmark study done in 1987 for the Commission for Racial Justice of the United Church of Christ indicates that three out of every five black and Hispanic Americans live in a community of one or more hazardous waste sites, landfills, incinerators, and polluting industries (Commission for Racial Justice 1987).

6. Compared to Europeans, Americans lived in sparsely populated smaller towns. In 1820 fewer than 7 percent of the population lived in cities and until 1910, 62 percent of Americans still lived in rural areas (Melosi 2000, 22).

7. The unfortunate events of September 11, 2001, have renewed the potency of the issue, as the towers were reduced to a pile of rubble that was hauled and buried in Fresh Kills landfill on Staten Island.

8. The Sanitation Department named Ukeles Fresh Kills' official artist in 1989. Since then, she has participated in and overseen various plans and projects turning the landfill's gigantic size (2,400 acres) and multiple facets into programmatic and aesthetic material for transformation. Following the recent conclusion of *Fresh Kills: Landfill to Landscape,* a design competition seeking a vision and a master plan for the site, and pending contract negotiations with the winning design teams (Field Operations led by James Corner and Stan Allen or the second and third place winners), Ukeles will collaborate with the chosen team (Krinke 2002).

9. See the EPA web site, www.epa.gov/epaoswer/non-hw/muncpl/landfill/index.htm (accessed 11/28/01).

10. In the early 1990s, researchers at Rutgers University tested various plant species at Fresh Kills for growth and tolerance. They found that six-year-old trees developed a shallow root system that did not penetrate the landfill cap; thus, these species can safely be planted on former dumps (Young 2001).

11. A central element in Ukeles's work is innovation in material reuse. For

the Danehy Park project, she performed several tests for the reuse of glass and rubber in paving and had the city approve the results before implementation.

12. The HMDC decided that since the design is based on accurate alignments, "further settlement of the landfill is necessary before [the project] can proceed." However, Nancy Holt, who was well aware of the settlement factor when she designed *Sky Mound,* believes that "since the star, Sun, and Moon alignments are all to the distant horizon, they will not be altered as the landfill settles" (1995, 62).

13. In the Netherlands, 75 percent of all waste is recycled, and the waste flow is too small to keep all the dumping sites profitable (Dooren 1999, 104).

14. Dooren mentions a demolition waste site near Lelystad in the Netherlands (1999, 105). In the United States, landfills in Collier County, Florida; Barre, Massachusetts; Bethlehem, New Hampshire; Edinburg, New York; and Lancaster, Pennsylvania, have been excavated. See the Landfill Mining web site, accessed at www.enviroalternatives.com/landfill (6/14/01).

15. Hiriya landfill in Tel Aviv—a thirty-five-year-old dump covering 120 acres, rising 200 feet, and filled with over 40 million cubic yards of garbage and soil—was closed in 1998. Its visual prominence and its central location close to major highways and to the national airport make it Israel's most potent symbol of a consumer culture excessively reliant on throwaway goods and of environmental neglect. Bavel Landfill near Breda is less prominent—at 85 acres, 105 feet, and 8 million cubic yards of waste—but equally culturally charged.

16. The proposals, solicited from well-established, internationally renowned artists and from younger, lesser-known artists and designers (including myself) whose area of inquiry coincides with the subject of the exhibition, were scrutinized by a committee of respectable museum curators. Artists were given no specific program or limitations. There was no obligation or funding to realize any of the proposals.

17. The Israeli Ministry of Interior and Ministry of Environmental Protection, which are responsible for future plans for the site, have recently embraced the idea of creating a recycling park around the garbage mountain.

CHAPTER 4. Waste Recycling / Reuse Institutions

1. In what follows, I do not discuss how to improve community recycling programs or promote recycling but rather interpret the phenomena of recycling and its host places, examining their opportunities for healthy, productive, and vital landscapes.

2. In the year 2000 the national recycling rate was 32 percent compared to 8 percent in 1990; 133.2 million Americans (61 percent of the population) receive curbside service (Goldstein and Madtes 2000, 40).

3. About four hundred incinerators were operating in the United States and Canada by 1914; only forty-five reduction plants were established at the time (Strasser 1999, 135).

4. Oregon was the first state to pass the "bottle bill" for can and bottle refund in 1972.

5. By 2000, nationwide there were 3,800 private yard trimmings composting facilities, mostly located near operating dumps and transfer stations or at closed dumps locations (Goldstein and Madtes 2000, 41).

6. WTE plants use hot gases from the burning process to produce steam, which is used in turn to produce electricity.

7. An example of an eco-industrial site is found in Kalunborg, Denmark. See first textbook on the subject by Graedel and Allenby, *Industrial Ecology* (1995).

8. The Salvation Army was founded in 1865 in London by William Booth to provide employment for the unskilled poor and to collect unwanted goods (Hoy and Robinson 1979, 1).

9. The MFA is a component of the New York City Department of Cultural Affairs and a partner of the Department of Sanitation's Bureau of Waste Prevention, promoting the city's waste prevention and reuse efforts (Heumann 1998).

10. Acceptable practices still functioning in various developed countries and facilitating the flow of items to the junk market are scarce. The *Alte Zachen* (a Yiddish idiom, meaning "old things") dealer still rides his horse-drawn cart every day through the streets of Tel Aviv, Israel, loudly announcing his presence via an amplifier, ready to pay small sums for household reusables. In Japan, members of a ragpickers' community called Aunt's Villa, located on a closed dump on the waterfront of Tokyo Bay, have demonstrated a unique ability to organize and transform their salvage activity into a respectable cooperative (Taira 1972).

11. Future plans for the closed landfill and its surroundings include a nature reserve.

12. The artist Steven Antonakos adorned the structure with a sculpture of colored neon lights and windows that glow orange at dusk.

13. The cost of the project ended up too high, and the project initiators were not able to secure a company for future acquisition and transfer.

14. Although the Wellesley facility is now operated by the municipality, it is included in this category of projects because it was initiated by grassroots organization and imbued with a small town community atmosphere.

CHAPTER 5. Sewage Treatment Plants

1. Throughout this chapter, many references are made to the Parisian sewer, as French writings on the subject are rich and extensive. They treated

the sewer metaphorically, holistically, and sensually, as well as practically. Although the English were also preoccupied with the subject, their attitudes and writings were quite pragmatic, deodorized, and genteel by comparison.

2. Edwin Chadwick blamed inadequate sanitation for the diseases and poverty that ravaged London's laborers, and he made many proposals for handling sewage (Chadwick 1965). Also see Samuel Edward Finer, *The Life and Times of Sir Edwin Chadwick* (1952).

3. Secondary treatment is the minimum level required in the United States and other developed countries. The tertiary process can remove more than 99 percent of all impurities from sewage (as opposed to only 60 percent in the primary phase and 85 percent in the secondary phase), producing an effluent that is almost potable. Because tertiary treatment can be very expensive and space-consuming, it is used only when the receiving water body is a fragile ecosystem.

4. Stallybrass and White explain that rats were initially an economic liability—pests that destroyed grains and other valuables. Later they turned symbolic, became a threat to civilized life, and were viewed through the lens of the century's sanitary and medical developments. During the grand sewer projects of London and Paris the rat became a source of fear, a demonized Other, and stories of rats attacking humans increased. Like the sewer itself, rats were a source of both horror and fascination in the bourgeois social imagination (1986, 143).

5. Montfaucon generated a half-million francs' worth of dry fertilizer per year (Reid 1991, 72).

6. In England and Scotland intense debates over and experiments to test the benefits to agriculture yield persisted throughout the century. In 1857 a Royal Commission on the Sewage of Towns was appointed to examine the benefits and limits of sewage farming (Sheail 1996).

7. Dewatered sludge, which was used as a fertilizer for devastated land in America and around the world in the early 1980s, is also made into commercial compost or dry pellets that can be burned to produce heat.

8. All definitions quoted herein are from *OED*, 1989 unless noted otherwise.

9. In 1999 the EPA estimated the investment necessary for wastewater systems at roughly $140 billion over the next twenty years, including wastewater treatment, upgrades of existing collection systems, new sewer construction, and control of sewer overflow. Two thousand additional plants could be necessary by 2016 to meet expanded treatment goals (Turner 1999, 10).

10. The artist Buster Simpson was recently commissioned to work on the public art component of the project.

11. For example, the approach taken to the aesthetics of the six new sludge dewatering plants built in the 1990s in New York City, designed by the ar-

chitect Richard Dattner, varied according to their location. Facilities in Oakwood Beach on Staten Island and in Bowery Bay in Queens are adorned with sculptural details and colorful wall panels, whereas the plants in Jamaica Bay, Queens, and Hunts Point in the Bronx that are closer to an airport and to industrial zones are simple and plain (Dattner 1995).

12. The treatment process leaves the water relatively clean and the sludge very toxic and dirty. Nitrogen, a useful plant growth matter, is converted to gases.

13. For map and details of the cloaca, see Katherine Rinne, *"Aqua Urbis Romae: The Waters of the City of Rome,"* November 30, 2001, accessed from www.iath.virginia.edu/waters (12/1/01).

14. Many of the twenty-four dumps sited around Paris in the 1880s manufactured ammonium sulfate, which fouled the air with fumes.

15. Brooklyn built its sewers in 1855, Chicago in 1856, and Jersey City in 1859 (Tarr 1988, 166).

16. A century later Chicago built a series of aeration pump stations (sidestream elevated pool aeration pumps, or SEPAs) to replenish the canal with oxygen, creating at the same time five neighborhood parks, ranging from 2 to 20 acres, whose centerpiece is a dramatic waterfall, the aeration mechanism itself (Robison 1994).

17. For example, in the 1930s New York City dumped its sludge 10 miles off the coast; by 1980s that distance had been extended to 100 miles. Ocean dumping was banned in 1992.

18. In their inclusive annotated bibliography, Suellen Hoy and Michael Robinson conclude that "Few Areas of public works history have received less attention from scholars than urban wastewater systems. Therefore many of the sources cited in their bibliography are dated books and articles largely intended for engineering audiences. Only in the 1970s, in conjunction with state and federal anti-pollution laws, did historical academic accounts appear and the public informed" (1982, 165).

19. New York alone built five of its current fourteen sewage plants between 1935 and 1937 and planned three others during the same period.

20. The 1948 Water Pollution Control Act passed by Congress created a federal authority to aid in building sewers and monitoring water pollution. In 1956, and then again in 1961, amendments to the law provided more impetus to control pollution. System expansion and new treatment plants were targeted again during the 1960s. The 1965 Water Quality Act and the 1972 Clean Water Act were critical for new sewage plant constructions.

21. The artist Ilya Kabakov staged shared toilets of Russian communal living quarters and played a melancholy melody behind a shit smeared door, signifying the toilet as a refuge from the everyday Soviet life. Sewer videos by Christian Boltansky showed swimming objects inside sewer pipes, and a film

of a sewer monitor camera by Peter Fischli and David Weiss revealed the vulgar and banal inner space of the pipes that could not be accessed otherwise.

22. These designs followed the elegant architecture of some turn-of-the-century European precedents, such as the sewage pump house in Abbey Mills, east of London, and that of Duisburg-Beeck in Duisburg, Germany.

23. Two examples of such industrial elegance are the Lynetten and Damhusaen sewage plant near Copenhagen, Denmark, and Matsushima Aizu Sewer Plant Control Facility, Amakusa, Japan. See *Arkitectur Denmark* 41, no. 3 (1997).

24. Recently built sewage pumps in Kings County, Washington—the North Creek Sewage Pump Station and the Interurban Pump Station near Tukwila—were embellished with artwork.

25. The engineers' proposal to build a conspicuous, two-story structure that would have blocked views of San Francisco Bay was successfully resisted by local residents and environmentalists.

26. The culvert was also intended to act as a resonating chamber, a kind of a water clock. Fountains, wetland oases, and riprap-filled gabions metaphorically and literally clean the water and accent some of its siltation and tidal processes (Simpson 1994).

27. An earlier conceptual proposal by the artist Haim Steinbach, who lives in the neighborhood, proposed a grove of trees atop an elevated terrace of serpentine retaining walls along the periphery of the plant (Public Art Fund 1993, 25).

28. Vancouver, Washington, treats other public utility facilities as public parks. Its water treatment plant is in the middle of a park that also contains an amphitheater for concerts and other major events.

29. Wolverton developed these systems in the mid-1970s as part of a project to study "closed ecological life support systems," a sustainable unit for space travel. Other pioneer researchers in the field were John Todd of Ocean Arks Institute in Burlington, Vermont, and Sola Aquasystem of Encinitas, California.

30. Other artists have since begun employing biological wastewater treatment in their designs. Christy Rupp integrated a wetland in her proposal for New York's Coney Island plant. It was to be constructed at the water's edge along the promenade and act as an extension of the functions of the plant itself, polishing the effluent (Cembalest 1991; Cameron 1994).

31. Jordan was recently hired to use the 2 percent art budget for the seaside Ventura Water Reclamation Plant near Ventura Harbor in California. She plans to direct her effort to improving the avian habitat and water quality.

32. The cost of the wetlands will be around $70 million, compared with the $630 million needed for a conventional facility upgrade.

33. New Zealand has been constructing marshes in a significant step to

protect its coastal and marine resources. In the United Kingdom, since the mid-1980s there has been a rapid growth in the use of wetlands and reed beds in combination with shallow oxidation ponds to improve effluent quality.

34. By 1998 the institute's trademarked Living Machines were operating in thirteen states in the United States and in seven other countries around the world (Lerner 1998).

35. At the Intervale Living Machine, a research and demonstration facility of the institute in Burlington, Vermont, food scraps and high-protein microbial communities are fed to tilapia, a freshwater fish grown in tanks. Fish waste dissolved in the water is then fed to vegetable and flowers. Worms grown on food scraps also serve as fish food, and their castings are used as fertilizer. See "Ocean Arks International," accessed at www.oceanarks.org (12/1/01).

36. See "Adam Joseph Lewis Center for Environmental Studies: The Lewis Center Living Machine," accessed at www.oberlin.edu/newserv/esc/living machine.html (12/1/01).

37. Viet Ngo owns the St. Paul–based Lemna Corporation. Ngo has already built 50 new plants and upgraded and rehabilitated about 125 systems.

38. Ngo hoped to build a lavatory with a room that would darken and project an image of the duckweed canal onto the wall as the visitor sits down to contribute waste, though the plan did not mature (Miller 1990; Kastner 1991).

39. Another plant, the Azolla fern, has been at the center of wastewater research; it is also able to reclaim from wastewater precious metals like gold, platinum, and silver and to detoxify hazardous metals, like uranium as it absorbs and bind them. The metals can later be recovered when the plant is incinerated.

EPILOGUE. Challenges for Thought and Action

1. This and the next quote are from "Fresno Landfill Archive," *Envirotech Newsletter,* accessed at www.geocities.com/erech/ (12/1/01); both messages were posted on August 31, 2001.

2. Forty thousand tons of waste at 131 locations in 39 states awaits burial. Two thousand tons are produced every day. See www.thebombproject.org (accessed 8/1/01).

3. Congress approved the storage of 77,000 tons of radioactive waste for more than 10,000 years. Target date for the facility opening is 2010.

4. See "Plutonium, the Contest: Rules," accessed at www.thebulletin.org/contest/rules.html (12/1/01). For descriptions of winning entries see *The Bulletin of Atomic Scientists,* May/June 2002.

References

Acconci Studio. 1997. "Preliminary Proposal #1 for Newtown Creek Wastewater Management Plant." Artist statement, April.

———. 1999. "Garbage City." In *Hiriya in the Museum,* edited by Martin Weyl, 50–55. Tel Aviv, Israel: Tel Aviv Museum of Art.

Ackerman, Frank. 1997. *Why Do We Recycle? Markets, Values, and Public Policy.* Washington, D.C.: Island Press.

Agnes, Denes. 1993. "Notes on Eco-Logic: Environmental Artwork, Visual Philosophy and Global Perspective." *Leonardo* 26, no. 5: 387–95.

Aldama, Frederick Luis, and Robert Soza. 2001. "Bad Subjects." *Political Education for Everyday Life* no. 55 (May). Accessed from http://eserver.org/bs/55/editors.html (4/25/02).

Alexander, Charles P. 1999. "What Would a Green Future Look Like?" *Time,* November 8, 115.

Alexander, Judd. 1993. *In Defense of Garbage.* Wesport, Conn.: Praeger.

Amato, Ivan. 1999. "Can We Make Garbage Disappear?" *Time,* November 8: 116.

Ammons, A. R. 1993. *Garbage.* New York: Norton.

APWA (American Public Works Association). 1997. "New Sewage Treatment Plant Transformed into Natural Resource Center That Is Community's Pride." *APWA Reporter* (May): 16.

Augé, Marc. 1995. *Non-Places: Introduction to an Anthropology of Supermodernity.* Translated by John Howe. New York: Verso.

Babias, Marius. 1994. "Cloaca Maxima." *Kunstforum International* no. 128 (October/December): 361–63.

Baggett, Rebecca. 1992. "The Art of Recycling: A Reuse Renaissance." *World Wastes* 35, no. 8: 33–36.

Bakhtin, Mikhail. 1984. *Rabelais and His World.* Translated by H. Iswolsky. Bloomington: Indiana University Press.

———. 1994. *The Bakhtin Reader.* Edited by Pam Morris. New London, Conn.: Edward Arnold.

Barlow, Ronald. 1989. *The Vanishing American Outhouse.* El Cajon, Calif.: Windmill Publishing.

Baudelaire, Charles. 1965. *The Painter of Modern Life, and Other Essays.* Translated and edited by Jonathan Mayne. London: Phaidon.

Beardsley, John. 1998. "Postnuclear Wilderness." *Landscape Architecture* 88, no. 4: 143–44.

Beecher, Catharine. 1843. *A Treatise on Domestic Economy.* Boston: Webb.

———. 1848. *A Treatise on Domestic Economy.* New York: Harper and Brothers.

———. 1873. *The New Housekeeper's Manual.* New York: Ford.

Benjamin, Walter. 1973. *Charles Baudelaire: A Lyric Poet in the Era of High Capitalism.* Translated by Harry Zohn. London: NLB.

Berger, John. 1991. "A Load of Shit." In *Keeping a Rendezvous,* 37–42. New York: Vintage.

BioCycle. 1994. "Industry News." *BioCycle* 35, no. 9: 87.

Blake, Peter. 1979. *God's Own Junkyard: The Planned Deterioration of America's Landscape.* New York: Holt, Rinehart, and Winston.

Block, David, and Annette Wood. 1998. "Rescuing Materials from Landfills." *BioCycle* 39, no. 2: 66–68.

Blumberg, Louis, and Robert Gottlieb. 1989. *War on Waste: Can America Win Its Battle with Garbage?* Washington, D.C.: Island Press.

Boorstin, Daniel. 1973. "Consumption Communities." In *The Americans: The Democratic Experience,* 89–160. New York: Random House.

Brataas, Anne. 1990. "The Art of Sewage." *Saint Paul Pioneer Press Express,* October 25, C1.

Breen, Bill. 1990. "Landfills Are #1." *Garbage* 2, no. 5: 42–45.

———. 1991. "Visionaries: The Future of Garbage." *Garbage* 3, no. 5: 26–35.

———. 1992a. "Garbage Dictionary" *Garbage* 4, no. 1: 13.

———. 1992b. "A Mountain and a Mission: Finding a Final Resting Place for Nuclear Garbage." *Garbage* 4, no. 3: 52–57.

BRING Recycling. 2001. "BRING's Grand Scheme Case Statement." Pamphlet, May 9.

Bronx Community Paper Company. 1997. "The History of the Bronx Community Paper Company." Accessed from www.bronxpaper.org/origins.html (9/25/99).

Brown, Patricia Leigh. 1989. "Space for Trash: A New Design Frontier." *New York Times,* July 27, C1.

Brown, William. 1851. *The Carpenter's Assistant.* Worcester, Mass.: Livemore.

Bruegmann, Robert. 1993. "Infrastructure Reconstructed." *Design Quarterly* 158, no. 4: 7–13.

Buck, Louisa. 1998. "Openings: Tomoko Takahashi." *Artforum* 36, no. 10: 120–21.

Bushman, Richard, and Claudia Bushman. 1988. "The Early History of Cleanliness in America." *Journal of American History* 74, no. 4: 1213–38.

Callahan, Sean. 2001. "Ken Dunn." *City Talk,* December 1, 5–6.

Calvino, Italo. 1974. *Invisible Cities.* Translated by William Weaver. New York: Harcourt Brace Jovanovich.

Cameron, Mindy Lehrman. 1994. "Rising above Our Garbage." Symposium report. The Exploratorium, San Francisco, Calif., January 28–30, 1993.

Canadian Centre for Architecture. 1991. *The Architecture of the Sanitation Movement 1870–1914*. Montreal: Canadian Centre for Architecture.

Canty, Donald. 1998. "Waterworks Garden." *Places* 12, no. 1: 17–19.

Caro, Robert A. 1974. *The Power Broker: Robert Moses and the Fall of New York*. New York: Knopf.

Cembalest, Robin. 1991. "Sheepshead Bay: A Sewage Plant Grows in Brooklyn." *Artnews* 90, no. 7: 55–56.

Certeau, Michel de. 1984. *The Practice of Everyday Life*. Translated by Steven R. Rendall. Berkeley: University of California Press.

Chadwick, Edward. 1965 [1842]. *Report on the Sanitary Condition of the Labouring Population of Great Britain*. Edited by M. W. Flinn. Edinburgh: University Press.

Chapman, Tony. 1999. "An Ideal Home, But Could You Really Live in It?" *New Statesman* 128: 32–33.

Chase, John. 1999. "A Curmudgeon's Guide to the Wild World of Trash." In *Everyday Urbanism*, edited by John Chase, Margaret Crawford, and John Kaliski, 52–67. New York: Monacelli Press.

Child, Georgie. 1925 [1914]. *The Efficient Kitchen*. New York: McBride.

Chu, Petra ten-Doesschate. 1993. "Scatology and the Realist Aesthetic." *Art Journal* 52, no. 3: 41–47.

City of Los Angeles Public Works. 1995. "Donald C. Tillman Water Reclamation Plant." Brochure, September.

City of Portland Bureau of Environmental Services. 1993. "Columbia Boulevard Wastewater Treatment Plant Headworks replacement Design. Project no. 4958. Design Memorandum no. 8: Landscape Design," March.

Classen, Constance, David Howes, and Anthony Synnott. 1994. *Aroma: The Cultural History of Smell*. New York: Routledge.

Clay, Grady. 1973. "Sinks." In *Close-Up: How to Read the American City*, 141–52. New York: Praeger.

———. 1994. "Out There," "Power Vacuum." In *Real Places: An Unconventional Guide to America's Generic Landscape*, 171–95. Chicago: University of Chicago Press.

Clendinen, Dudley. 1985. "A Garbage Dump That Is a Town Showpiece." *New York Times*, September 2, A7.

Clifton, Leigh Ann. 1993. "A Conversation with Lewis Buster Simpson." *Artweek* 24, no. 3: 24.

Coakley, Michael. 1986. "In Wellesley, People Wouldn't Dream of Trashing the Dump." *Chicago Tribune*, June 24, A18.

Commission for Racial Justice. 1987. "Toxic Wastes and Race in the United States: A National Report on the Racial and Socio-Economic Characteristics of Communities with Hazardous Waste Sites." Commission for Racial Justice, United Church of Christ. New York: Public Data Access.

Corbin, Alain. 1986. *The Foul and the Fragrant: Odor and the French Social Imagination.* Cambridge, Mass.: Harvard University Press.

Crewe, Jonathan. 1991. "Defining Marginality?" *Tulsa Studies in Women's Literature* 10, no. 1: 121–30.

Crossen, Cynthia. 1990. "A Sad Pall Hangs over the Social Hub of Nantucket Island." *Wall Street Journal,* August 22, A1.

Daniel, Julie. 2001. "Manager's Corner." *BRING Recycling's Used News* 10, no. 1: 2.

Dante Alighieri. 1954. *Inferno.* Translated by John Aitken Carlyle. London: J. M. Dent.

Darier, Eric. 1996. "The Politics and Power Effects of Garbage Recycling in Halifax, Canada." *Local Environment* 1, no. 1: 63–86.

Dattner, Richard. 1995. *Civil Architecture: The New Public Infrastructure.* New York: McGraw- Hill.

Dawson, Kerry. 1990. "Rehabilitating the Earth, San Francisco Style." *Landscape Architecture* 80, no. 12: 60–61.

DeLillo, Don. 1997. *Underworld.* New York: Scribner.

Del Porto, David, and Carol Steinfeld. 1999. *The Composting Toilet System Book.* Concord, Mass.: Center for Ecological Pollution Prevention.

Dooren, Noel van. 1999. "Never Again Will the Heap Lie in Peace." In *Tales of the Tip,* edited by Chris Driessen and Hiedi van Mierlo, 100–105. Amsterdam: Fundament Foundation.

Doran, Anne. 1996. "Flow City." *Grand Street* 57 (summer): 199–213.

Douglas, Mary. 1966. *Purity and Danger: An Analysis of Concepts of Pollution and Taboo.* New York: Praeger.

Doveil, Frida. 1997. "Eco-Performing Materials: Research, Tradition and Culture." *Domus* no. 789: 51–54.

Downing, Andrew Jackson. 1967 [1842]. *Cottage Residences, Rural Architecture and Landscape Gardening.* Watkins Glen, N.Y.: Library of Victorian Culture.

Drabelle, Dennis. 1990. "Sky Mound." In *The Art of Landscape Architecture,* 18–23. Washington, D.C.: Partners for Livable Places.

Driessen, Chris, and Heidi van Mierlo, eds. 1999. *Tales of the Tip: Art on Garbage.* Exhibition catalog. Amsterdam: Fundament Foundation.

Duncan, Michael. 1995. "Transient Monuments." *Art in America* 83, no. 4: 79–82.

Eliot, T. S. 1949 [1922]. *The Waste Land and Other Poems.* London: Faber and Faber.

Engler, Mira. 1995. "Waste Landscape: Permissible Metaphors in Landscape Architecture." *Landscape Journal* 14, no. 1: 10–25.

———. 1997. "Repulsive Matter: Landscapes of Waste in the American Middle Class Residential Domain." *Landscape Journal* 16, no. 1: 60–79.

———. 1999. "Re-Claiming Metaphors Out of the Dump, or Four Gestures for

Hiriya." In *Hiriya in the Museum,* edited by Martin Weyl, 44–49. Tel Aviv: Tel Aviv Museum of Art.

Engler, Mira, and Gina Crandell. 1993. "Open Waste System Park."In *The Once and Future Park,* edited by Deborah Karasove and Steve Waryan, 49. New York: Princeton Architectural Press.

Enzensberger, Christian. 1972. *Smut: An Anatomy of Dirt.* Translated by Sandra Morris. London: Calder & Boyars.

Erikson, Kai. 1994. "Out of Sight, Out of Our Mind." *New York Times Magazine,* March 6, 34–42.

Farre, Molly. 1997. "Inner City Recycling and Composting." *BioCycle* 38, no. 4: 30–32.

Ferguson, Russell. 1990. "Introduction: Invisible Center." In *Out There: Marginalization and Contemporary Cultures,* edited by Russell Ferguson et al., 9–14. New York: New Museum of Contemporary Art; Cambridge, Mass.: MIT Press.

Finer, Samuel Edward. 1952. *The Life and Times of Sir Edwin Chadwick.* London: Methuen.

Fisher, Risa and Bill Angelo. 1999. "Not Just a Big Green Hill." *Waste Age* 30, no. 1: 36–43.

Fisk, Pliny. 1993. "A Comparative Analysis and Recommendations for Alternative On- or Near-Site Wastewater Treatment Systems for the U.S.–Mexico Border." Center for Maximum Potential Building Systems, Inc. Unpublished report, January.

Foucault, Michel. 1986. "Of Other Spaces." Translated by Jay Miskwiec. *Diacritics* 6, no. 1 (spring): 22–27.

———. 1997 [1985]. "Of Other Spaces: Utopias and Heterotopias." In *Rethinking Architecture: A Reader in Cultural Theory,* edited by Neil Leach, 350–56. New York: Routledge.

Freud, Sigmund. 1961. *Civilization and Its Discontents.* Translated and edited by James Strachy. New York: Norton.

———. 1977. *On Sexuality: Three Essays on the Theory of Sexuality and Other Works.* Translated by James Strachey. Edited by Angela Richards. London: Penguin Books.

Gablik, Suzi. 1989. "Deconstructing Aesthetics." *New Art Examiner* 16, no. 5: 32–35.

Galfetti, Gustau Gili. 1999. "Das Paradies: Cornelius Kolig." In *My Home, My Paradise: The Construction of the Ideal Domestic Universe,* 182–89. Corte Madera, Calif.: Ginko Press.

Gantenbein, Douglas. 1995. "West Point Sewage Treatment Plant." *Architectural Record* 185: 86–89.

Gardner, Gary. 1998. "Fertile Ground or Toxic Legacy." *World Watch* 11, no. 1 (January/February): 28–34.

Gaulkin, Zachary. 1990. "Lush New Park Opens, Douses Memory of Landfill." *Cambridge Chronicle,* September 20.

Georgescu-Roegen, Nicholas. 1971. *The Entropy Law and the Economic Process.* Cambridge, Mass.: Harvard University Press.

Gerhard, Deborah. 1996. "Moralistic Eco-Fundamentalism: A Case Study of Byxbee Park." *Critiques of Built Works of Landscape Architecture.* Vol. 3. Baton Rouge: LSU School of Landscape Architecture, 21–23.

Giedion, Siegfried. 1948. *Mechanization Takes Commend: A Contribution to Anonymous History.* New York: Oxford University Press.

Girling, Cynthia and Kenneth Helphand. 1994. *Yard, Street, Park: The Design of Suburban Open Space.* New York: John Wiley.

Glatt, Linnea, Michael Singer, Laurel McSherry, and Frederick Steiner. 2000. "Center for Environmental Learning and Enterprise Master Plan." Report for the Herberger Center for Design Excellence, Arizona State University, Tempe, Arizona.

Goin, Peter. 1991. *Nuclear Landscapes.* Baltimore: Johns Hopkins University Press.

Goldbeck, David. 1989. *The Smart Kitchen: How to Create a Comfortable, Safe, Energy- Efficient, and Environmental-Friendly Workspace.* Woodstock, N.Y.: Ceres Press.

Goldstein, Nora, and Celeste Madtes. 2000. "The State of Garbage in America." *BioCycle* 41, no. 11: 40–48, 79.

Gora, Monika, and Gunilla Bandolin. 1996. "The Museum of Garbage." *Topos* 14, no. 3: 66–70.

Graedel, Thomas E., and Branden R. Allenby. 1995. *Industrial Ecology.* Englewood Cliffs, N.J.: Prentice-Hall.

Greene, Jan. 1993. "Investor Creates ATM for Recyclers." *Garbage* 4, no. 6: 20.

Greenfield, Venri. 1986. *Making Do or Making Art: A Study of American Recycling.* Ann Arbor, Mich.: UMI Research Press.

Gregoire, Mathieu. 1996. *Report on Landscape, Architectural and Aesthetic Improvements to the Point Loma Wastewater Treatment Plant.* Report prepared for the City of San Diego Metropolitan Wastewater Department and Commission for Arts and Culture, May 6.

Griffin, Vern. 1987. "Renowned Architect Has Designed Trash Facility." *San Diego Tribune,* September 17, B9.

Griswold, Mac. 1993. "The Landfill's Progress." *Landscape Architecture* 83, no. 10: 79–81.

Hanson, David, et al. 1997. *Waste Land: Meditations on a Ravaged Landscape.* New York: Aperture.

Hardin, Garrett. 1977. "The Tragedy of the Commons." In *Managing the Commons,* edited by Garrett Hardin and John Baden, 16–30. San Francisco: W. H. Freeman and Company.

Hargreaves Associates. 1996. "Parque do Tejo e Trancão." In *Landscape Transformed*, edited by Michael Spens, 90–93. London: Academy Editions.

———. 1997. "Saint-Michel Environmental Complex Landfill Park Development Plan." Unpublished report, January 27.

Harper, Brian Phillip. 1994. *Framing the Margins: The Social Logic of Postmodern Culture*. New York: Oxford University Press.

Hartney, Eleanor. 1990. "Garbage Out Front: A Review." *The Livable City* 14, no. 2: 13.

Herrmann, Gretchen, and Stephen Soiffer. 1984. "For Fun and Profit: An Analysis of the American Garage Sale." *Urban Life* 12, no. 4: 397–421.

Hess, Alan. 1992. "Technology Exposed." *Landscape Architecture* 82, no. 5: 39–49.

Heumann, Jenny M. 1998. "Trash to Art." *Waste Age* 29, no. 2: 57–59.

Hinte, Ed van, and Conny Bakker. 1999. *Trespassers: Inspirations for Eco-Efficient Design*. Netherlands Design Institute. Rotterdam: 010 Publishers.

HMDC (Hackensack Meadowlands Development Commission). 1992. "Richard W. DeKorte Park." Fact sheets, HMDC Environment Center Museum. Lyndhurst, N.J.

Holt, Nancy. 1995. "Nancy Holt." In *Sculpting with the Environment: A Natural Dialogue*, edited by Baile Oakes, 56–64.New York: Van Nostrand Reinhold.

Holtz, Kay Jane. 1990. "Landfill as Art Form: Going Beyond the Mound." *New York Times*, March 15, B1.

Hood, Walter. 1997. *Urban Diaries*. Washington, D.C.: Spacemaker Press.

Hooper, Louise. 1993. "Tale of a Tip." *Landscape Design* 219: 19–21.

Hopper, Joseph R., and Joyce McCarl Mielsen. 1991. "Recycling as Altruistic Behavior: Normative and Behavioral Strategies to Expand Participtation in a Community Recycling Program." *Environment and Behavior* 23, no. 2: 195–220.

Horan, Julie L. 1996. *The Porcelain God: A Social History of the Toilet*. Secaucus, N.J.: Crol.

Horning, Ron. 1987. "In Time: Earthworks, Photodocuments, and Robert Smithson's Buried Shed." *Aperture* 106, no. 1: 74–77.

Hoy, Suellen, and Michael Robinson. 1979. *Recovering the Past: A Handbook of Community Recycling Programs, 1890–1945*. Chicago: Public Works Historical Society.

———. 1982. *Public Works History in the United States: A Guide to the Literature*. Nashville: American Association for State and Local History.

Hugo, Victor. 1998 [1862]. *Les Misérables*. Translated by Norman Denny. New York: Penguin Books.

Hung, Su-Chen. 1997. "Water Spells." Artist statement.

Ibelings, Hans. 1998. *Supermodernism: Architecture in the Age of Globalization*. Rotterdam: NAi Publishers.

Illich, Ivan. 1985. *H2O and the Waters of Forgetfulness: Reflections on the Historicity of "Stuff."* Dallas: Dallas Institute of Humanities and Culture.

Interior Design. 1994. "The Bathroom: Evolution of the Species." *Interior Design* 55, no. 6: 228–31.

Iovine, Julie V. 1988. "Good Architecture in New York at Last." *The Connoisseur* 218, no. 914: 119–21.

Isaak, Jo Anna. 2001. "Trash: Public Art by the Garbage Girls." In *Gendering Landscape Art,* edited by Steven Adams and Anna Gruetsner Robins, 173–85. New Brunswick, N.J.: Rutgers University Press.

Jackson, J. B. 1980. "The Domestication of the Garage." In *The Necessity for Ruins, and Other Topics,* 103–12. Amherst: University of Massachusetts Press.

Jakle, John, and David Wilson. 1992. *Derelict Landscapes: The Wasting of America's Built Environment.* Savage, Md.: Rowman and Littlefield.

Jenner, Mark. 1997. "Underground, Overground: Pollution and Place in Urban History." *Journal of Urban History* 24, no. 11 (November): 97–110.

Johanson, Patricia. 1991. *Patricia Johanson: Public Landscapes.* Philadelphia: Painted Bride Art Center.

———. 1992. *Art and Survival: Creative Solutions to Environmental Problems.* North Vancouver, B.C.: Gallerie Publications.

———. 1993. "Art and Ecology: A Precarious Balance." Speech to the annual conference of the College Arts Association, Seattle, Wash., February 5.

Johnson, H. C. 1969. "Gravel+Refuse=Recreation." *Parks & Recreation: Official Publication of the National Recreation and Park Association* 4, no. 9: 46–48.

Johnson, Jory. 1996. "Country Living." *Landscape Architecture* 86, no. 2: 40.

Johnson, Ken. 2002. "A Landfill in the Eyes of Artists Who Beheld It." *New York Times,* February 1, E38.

Kalven, Jamie. 1991. "Trash Action." *University of Chicago Magazine,* April, 17–23.

Kastner, Jeffrey. 1991. "The Duckweed Factor." *Artnews* 90, no. 2: 37.

———. 1999. "Deep Time Design." *Public Art Review* 21: 11–13.

Keefer, Bob. 2001. "What a Dump!" *Register Guard* (Eugene, Oregon), January 21, 1H.

Kelly, Katie. 1973. *Garbage: The History and Future of Garbage in America.* New York: Saturday Review Press.

Kendrick, Laura. 1992. "The Monument and the Margin." *South Atlantic Quarterly* 91, no. 4: 835–64.

Kira, Alexander. 1976. *The Bathroom.* New York: Viking.

Kissida, John, and Nancy Beaton. 1991. "Landfill Park: From Eyesore to Asset." *Civil Engineering* 61, no. 8: 49–51.

Klein, Richard. 1995. "Get a Whiff of This: Breaking the Smell Barrier." *The New Republic,* February 6, 18–23.

Kleine, Helene. 1994. "Landmarks in the Emscher Park." *Topos* 11 (August): 97–99.

Knapp, Gottfried von. 1998. "Orte der Verdrängung." *Baumeister* 95, no. 3: 38–47.

Kourik, Robert. 1990. "Toilets: The Law-Flush/No Flush Story." *Garbage* 2, no. 1: 16–23.

———. 1992. "Even During Droughts Cisterns Deliver Water." *Garbage* 4, no. 4: 43–46.

———. 1995. "Graywater for Residential Irrigation." *Landscape Architecture* 85, no. 1: 30–33.

Krinke, Rebecca. 2002. "Fresh Ideas?" *Landscape Architecture* 92, no. 6: 76–85.

Kristeva, Julia. 1982. *Powers of Horror: An Essay on Abjection.* Translated by Leon S. Roudiez. New York: Columbia University Press.

Lacan, Jacques. 1968. *The Language of the Self: The Function of Language in Pyschoanalaysis.* Edited by Anthony Wilden. New York: Dell.

Lahiji, Nadir, and D. S. Friedman. 1997. "At the Sink." In *Plumbing: Sounding Modern Architecture,* 35–55. New York: Princeton Architectural Press.

Lambert, Bruce. 1999. "A Garbage Heap Turns to Paradise; More Island Landfills Could Become Parks." *New York Times,* May 9, LI1.

Lambton, Lucinda. 1995. *Temples of Convenience and Chambers of Delight.* New York: St. Martin's.

Langdon, Phillip. 1994. *A Better Place to Live: Reshaping the American Suburb.* Amherst, Mass.: University of Massachusetts Press.

Laporte, Dominique. 2000. *History of Shit.* Translated by Nadia Benabid and Rodolphe el- Khoury. Cambridge, Mass.: MIT Press.

Leccese, Michael. 1995. "Plumbing the Heights." *Landscape Architecture* 85, no. 6: 55–57.

———. 1997. "Cleansing Art." *Landscape Architecture* 87, no. 1: 70–75.

———. 1999. "A Point Well Taken." *Landscape Architecture* 89, no. 6: 64–69.

———. 2001. "Enigma in the Garden." *Landscape Architecture* 91, no. 5: 75–82.

Lerner, Steve. 1997. "Pliny Fisk III: The Search for Low-Impact Building Materials and Techniques." In *Eco-Pioneers Practical Visionaries Solving Today's Environmental Problems,* 19–36. Cambridge, Mass.: MIT Press.

———. 1998. "Living Machines: The New Environmentalists." *The Futurist* 32, no. 4 (May): 35–40.

Leroux, Pierre. 1853. *Aux États de Jersey.* London: Universal Library.

Lifset, Reid. 1990. "Nimby's and the Politics of Garbage." *The Livable City* 14, no. 2) (November): 5–6.

Linebaugh, Donald. 1994. "All the Annoyances and Inconveniences of the Country." *Winterthur Portfolio* 29, no. 1: 1–18.

L'observatoire International. 1999. "Lighting Design for the Newtown Creek Sewage Plant." Project statement.

Loos, Adolf. 1997. "Plumbers." In *Plumbing: Sounding Modern Architecture,* edited by Nadir Lahiji and D. S. Friedman, 15–19. New York: Princeton Architectural Press.

Lovelace, Carey. 1987. "Sky Mound." *Arts New Jersey,* Fall: 10–14.

Lupton, Ellen, and Abbott Miller. 1992. *The Bathroom, the Kitchen, and the Aesthetics of Waste: A Process of Elimination.* New York: Kiosk.

Luton, Larry S. 1996. *The Politics of Garbage: A Community Perspective on Solid Waste Policy Making.* Pittsburgh: University of Pittsburgh Press.

Lynch, Kevin. 1972. "Managing Transitions." In *What Time Is This Place?,* 190–214. Cambridge, Mass.: MIT Press.

———. 1990. *Wasting Away.* Edited by Michael Southworth. San Francisco: Sierra Club Books.

MacCannell, Dean. 1976. *The Tourist: A New Theory of the Leisure Class.* New York: Schocken Books.

Marceca, Maria Luisa. 1981. "Reservoir, Circulation, Residue: JCA Alphand, Technological Beauty and the Green City," *Lotus International* 30, no. 1: 57–63.

Marinelli, Janet. 1990. "After the Flush: The Next Generation." *Garbage* 2, no. 1: 24–35.

Marter, Joan. 1994. "Nature Redux: Recent Reclamation Art." *Sculpture* 13, no. 6: 30–33.

Martin, Michael. 1995. "Learning From Alleys: The Back Alley in the American Residential Landscape." M.A. thesis, University of Oregon.

Marton, Deborah. 1996. "Hidden Opportunities." *Landscape Architecture* 86, no. 7: 38–43.

Masterson, Ikuno. 1981. "Cooperation in Waterfront Design." *Small Town* 12, no. 3: 53–55.

Matilsky, Barbara. 1992. *Fragile Ecologies: Contemporary Artists' Interpretations and Solutions.* New York: Rizzoli.

McDonald, Jackie. 1990. "Mining Recyclables from our Garbage." *Garbage* 2, no. 5: 16–18.

McEvilley, Thomas. 1999. "If It Die: Triangulating Landfill Art." In *Tales of the Tip,* edited by Chris Driessen and Hiedi van Mierlo, 60–63. Amsterdam: Fundament Foundation.

McLaughlin, Terence. 1988. *Dirt: A Social History as Seen Through the Uses and Abuses of Dirt.* New York: Dorset Press.

McShane, Clay. 1979. "Transforming the Use of Urban Space: A Look at the Revolution in Street Pavements, 1880–1924." *Journal of Urban History* 5, no. 3: 279–307.

McSherry, Laurel, and Frederick Steiner. 1998. "Urban Wilds: The 27th Avenue Solid Waste Management Campus Plan." *Recycling Review Newsletter* (Winter). Arizona State University: Herberger Center for Design Excellence and Center for Environmental Studies.

———. 2000. "Recovering an Urban Wasteland." *Plurimondi* 2, no. 3: 143–60.

Meinig, D. W., ed. 1979. *The Interpretation of Ordinary Landscapes: Geographical Essays*. New York: Oxford University Press.

Melnick, Mimi. 1994. *Manhole Covers*. Cambridge, Mass.: MIT Press.

Melosi, Martin. 1981. *Garbage in the Cities: Refuse, Reform, and the Environment, 1880–1980*. College Station: Texas A&M University Press.

———. 1982. "Battling Pollution in the Progressive Era." *Landscape* 26, no. 3: 35–41.

———. 2000. *The Sanitary City: Urban Infrastructure in America from Colonial Times to the Present*. Baltimore: Johns Hopkins University Press.

———, ed. 1980. *Pollution and Reform in American Cities, 1870–1930*. Austin: University of Texas Press.

Meyer, Elizabeth. 1991. "The Public Park as Avant-Garde (Landscape) Architecture." *Landscape Journal* 10, no. 1: 16–26.

Miller, Benjamin. 2000. *Fat of the Land: Garbage in New York City, The Last Two Hundred Years*. New York: Four Walls Eight Windows.

Miller, Cheryl. 1990. "Viet Ngo: Infrastructure as Art." *Public Art Review* 2, no. 1: 20–21.

Miller, Marc H. 1989. "Something for Everyone: Robert Moses and the Fair." In *Remembering the Future: The New York World's Fair from 1939 to 1964*, edited by the Queens Museum, 45–75. New York: Rizzoli.

Miller, Max H. 1988. "Patterns of Exchange in the Rural Sector: Flea Markets along the Highway." *Journal of American Culture* 11, no. 3: 55–59.

Mirach, Robert. 1990. *Bravo 20: The Bombing of the American West*. Baltimore: Johns Hopkins University Press.

Moggridge, Hal. 1993. "A Frank Artifact." *Landscape Design* 219: 22–23.

Molesworth, Helen. 1997. "Bathrooms and Kitchens: Cleaning House with Duchamp." In *Plumbing: Sounding Modern Architecture*, edited by Nadir Lahiji and D. S. Friedman, 75–92. New York: Princeton Architectural Press.

Mollison, Bill. 1990. *Permaculture: A Practical Guide for a Sustainable Future*. Washington, D.C.: Island Press.

Montaigne, Michel de. 1948. "Of Smells." *The Complete Essays of Montaigne*, translated Donald M. Frame, 228–29. Stanford, Calif.: Stanford University Press.

Morrell, Ricki. 1998. "Town's Great Passion Is But a Recycling of Its Puritanical Past" *Wall Street Journal*, February 4, NE1.

Morrissey, Lee. 2000. "Desolate Lands, Desolate Minds: The Therapeutic Function of the Project." *Plurimondi* 2, no. 3: 29–46.

Mumford, Lewis. 1961. *The City in History: Its Origins, Its Transformations and Its Prospects*. New York: Harcourt, Brace and World.

Muschamp, Herbert. 1993. "When Art Is a Public Spectacle." *New York Times,* August 29, B1.

————. 1994. "When Is a Roof Not a Roof?" *New York Times,* April 10, Arts and Leisure 42.

————. 1998. "Greening a South Bronx Brownfield." *New York Times,* January 23, E33.

Mydans, Seth. 2000. "A World of Scavengers on the Fringes of Wealth." *New York Times* July 23, WK15.

Nobel, Philip. 1999. "As Always, Please Touch." *New York Times,* April 8, F1.

O'Connell, Kim. 1998. "The Closure of Fresh Kills: What's the Forecast?" *Waste Age* 29, no. 10: 48–57.

————. 1999. "Out of the Shadow of Mount Royal." *Landscape Architecture* 89, no. 6: 44–50.

OED (Oxford English Dictionary). 1989. 2d ed. New York: Oxford University Press.

Ogle, Maureen. 1988. "Municipal Services in the Awakened City: Waste Collection and Disposal in America, 1900–1920." Unpublished paper.

———— 1993. "Domestic Reform and American Household Plumbing, 1840–1870." *Winterthur Portfolio* 28, no. 1: 33–58.

Olcott, John, and Tom Pedersen. 1985. "Visual Quality for Rapid Infiltration Basins." *Landscape Architecture* 75, no. 6: 55–57.

Oskamp, Stuart, et al. 1991. "Factors Influencing Household Recycling Behavior." *Environment and Behavior* 23, no. 4: 494–519.

Packard, Vance. 1960. *The Waste Makers*. New York: Simon and Schuster.

Palmer, Phyllis. 1989. *Domesticity and Dirt: Housewives and Domestic Servants in the United States, 1920–1945*. Philadelphia: Temple University Press.

Parent-Duchâtelet, Alexandre. 1837. *De la Prostitution dans la Ville de Paris*. 2 vols. 2d ed. Paris: J. B. Baillière.

Pawley, Martin. 1975. *Garbage Housing*. London: Architectural Press.

————. 1982. *Building for Tomorrow: Putting Waste to Work*. San Francisco: Sierra Club Books.

Pearson, Clifford, 1998. "A Dutch Waste Transfer Station Uses Architecture to Be a Good Neighbor." *Architectural Record* 186, no. 10: 102–3.

Petersen, Jon. 1979. "The Impact of Sanitary Reform upon American Urban Planning, 1840–1890." *Journal of Social History* 13, no. 1: 83–104.

Phillips, Patricia. 1989. "Public Art—Waste Not." *Art in America* 77, no. 2: 47–51.

————. 1990. "Recycling Metaphors: The Culture of Garbage." *The Livable City* 14, no. 2: 12.

Phoenix Arts Commission. 1991a. "23rd Ave. Wastewater Treatment Plant." Project fact sheet.

———. 1991b. "91st Avenue Wastewater Treatment Plant, 'Hydrotifer.'" Project fact sheet.

Plante, Ellen. 1995. *The American Kitchen 1700 to the Present: From Hearth to Highrise.* New York: Facts on File.

Pliny the Elder. 1957. *Pliny's Natural History: An Account by a Roman of What Romans Knew and Did and Valued.* Translated by Lloyd Haberly. New York: Frederick Ungar.

Pollard, Sidney. 1997. *Marginal Europe: The Contribution of Marginal Lands since the Middle Ages.* New York: Oxford University Press.

Poore, Jonathan. 1989. "Kitchen Design for Recycling," *Garbage* 1, no. 1: 18–24.

Poore, Patricia. 1990. "Towards the Next Kitchen." *Garbage* 2, no. 4: 35.

Potteiger, Matthew, and Jamie Purinton. 1998. "The Wasteland & Restorative Narrative." In *Landscape Narratives*, 213–39. New York: Van Nostrand Reinhold.

Powell, Kevin. 1992. "Topographic Statement." *Landscape Architecture* 82, no. 1: 37–39.

Primard, Philippe, and Yves Brangier. 1995. "A Landscape for the Achéres Sewage Plant." *Topos* 10, no. 3: 43–47.

Public Art Fund. 1993. *Urban Paradise.* Competition catalogue winning proposals. New York.

Rainey, Reuben. 1994. "Environmental Ethics and Park Design: A Case Study of Byxbee Park," *Journal of Garden History* 14, no. 3: 171–78.

Rathje, William. 1990. "The History of Garbage: Archeologists Bust Myths about Solid Waste and Society." *Garbage* 2, no. 5: 32–39.

Rathje, William, and Cullen Murphy. 1992. *Rubbish! The Archeology of Garbage.* New York: HarperCollins.

Reid, Donald. 1991. *Paris Sewers and Sewermen: Realities and Representations.* Cambridge, Mass.: Harvard University Press.

Reynolds, Reginald. 1943. *Cleanliness and Godliness.* London: Allen and Unwin.

Richardson, Benjamin Ward. 1876. *Hygeia: A City of Health.* London: Macmillan.

Rifkin, Jeremy. 1980. *Entropy: A New World View.* New York: Viking.

Riggle, David. 1994. "Creating Markets Close to Home". *BioCycle* 35, no. 7: 78–82.

———. 1995. "Integrated Processing At Future recycling Park," *BioCycle* 36, no. 10: 30–32.

———. 1996. "Technology Improves for Composting Toilets." *BioCycle* 37, no. 4: 39–43.

Ritter, Arno. 1997. "Cornelius Kolig: Pradise." *Domus* 794 (June): 44–51.

Robison, Rita. 1994. "Chicago's Waterfalls." *Civil Engineering* 64, no. 7: 36–39.

Rockefeller, Abby. 1996. "Civilization and Sludge: Notes on the History of the Management of Human Excreta." *Current World Leaders* 39, no. 6: 99–113.

Rosenberg, Harold. 1989. "Collage: Philosophy of Put-Togethers."In *Collage Critical Views,* edited by Katherine Hoffman, 59–64. Ann Arbor, Michigan: UMI Research Press.

Rudofsky, Bernard. 1984. "Uncleanliness and Ungodliness." *Interior Design* 55, no. 6: 212–21.

Schmertz, Mildred. 1994. "Ecological Control." *Architecture* 83, no. 8: 58–65.

Schrödinger, Erwin. 1967. *What Is Life?: The Physical Aspects of the Living Cell, with Mind and Matter.* New York: Cambridge, University Press.

Schultz, Stanely. 1989. *Constructing Urban Culture: American Cities and City Planning, 1800- 1920.* Philadelphia: Temple University Press.

Schuyler, David, and Jane Turner Censer, eds. 1992. *The Papers of Frederick Law Olmsted.* Vol.6. Baltimore: Johns Hopkins University Press.

Shabecoff, Philip. 1993. *A Fierce Green Fire: The American Environmental Movement.* New York: Hill and Wang.

Shapiro, Gary. 1988. "Entropy and Dialectic: The Signatures of Robert Smithson," *Arts Magazine* 62, no. 10: 99–104.

Sheail, John. 1996. "Town Wastes, Agricultural Sustainability and Victorian Sewage." *Urban History* 23, no. 2: 189–210.

Shields, Rob. 1991. *Places on The Margins: Alternative Geographies of Modernity.* London: Routledge.

———. 1992. "The Individual, Consumption Cultures and the Fate of Community." In *Lifestyle Shopping: The Subject of Consumption,* edited by Rob Shields, 99–113. London: Routledge.

Silk, Gerald. 1993. "Myth and Meaning in Manzoni's 'Merda d'artista'" *Art Journal* 52, no. 3: 65–75.

Simmel, Georg. 1950."The Stranger." *The Sociology of Georg Simmel,* edited and translated by Kurt H. Wolff, 402–8. New York: The Free Press.

Simmons, Elizabeth. 1993. "Means to Restore." *Landscape Design* 219: 15–18.

Simpson, Buster. 1994. "A Dialogue along the Danube Canal." Project report submitted to the City of Vienna, Austria.

———. 1998. "Poetic Utility." Seattle Public Utilities Arts master plan. Commissioned by the Seattle Arts Commission, April.

Smithson, Robert. 1979. *The Writings of Robert Smithson.* Edited by Nancy Holt. New York: New York University Press.

———. 1996. "Entropy Made Visible." In *Robert Smithson: The Collected Writings,* edited by Jack Flam, 301–8. Berkeley: University of California Press.

Soper, Kate. 1996. "Nature / 'nature.'" In *FutureNatural: Nature, Science, Culture,* edited by George Robertson et al., 22–34. New York: Routledge.

Sorvig, Kim. 2002. "Waste Not." *Landscape Architecture* 92, no. 7: 44–48.

Spiller, Larry. 1985. "Wastewater Oasis at Bishop's Lodge." *Landscape Architecture* 75, no. 6: 49–51.

Stallybrass, Peter, and Allon White. 1986. *The Politics and Poetics of Transgression.* London: Methuen.

Stegner, Wallace. 1962. "The Dump Ground." In *Wolf Willow: A History a Story and a Memory of the Last Plains Frontier,* 31–36. New York: Viking

Steuteville, Robert. 1994. "The State of Garbage in America." *BioCycle* 35, no. 4: 46–52.

Stevens, Kimberly. 1998. "Nantucket: What a Dump." *New York Times,* August 20, F4.

Stilgoe, John. 1988. *Borderland: Origins of the American Suburb, 1820–1939.* New Haven: Yale University Press.

Stone, May. 1979. "The Plumbing Paradox: American Attitudes toward Late Nineteenth Century Domestic Sanitary Arrangements," *Winterthur Portfolio* 14, no. 3: 283–309.

Strang, Gary. 1996. "Infrastructure as Landscape." *Places* 10, no. 3: 8–15.

Strasser, Susan. 1982. *Never Done: A History of American Housework.* New York: Pantheon.

———. 1992. *Waste and Want: The Other Side of Consumption.* German Historical Institute. Annual Lecture Series no. 5. Providence, R.I.: Berg Publishers.

———. 1999. *Waste and Want: A Social History of Trash.* New York: Metropolitan Books.

Sullivan, Robert. 1998. *The Meadowlands: Wilderness Adventures at the Edge of the City.* New York: Anchor Books.

Suneson, Torbjörn. 1999. "Sweden: Top Managers for Open Spaces." *Topos* 27 (June): 50–55.

Sunset Magazine. 1952. "Landscaping Your Home." Menlo Park, Calif.: Lane Publishing Co.

———. 1981. "Ideas for Kitchen Storage." Menlo Park, Calif.: Lane Publishing Co.

Sutro, Dirk. 1993. "Bravo 20: Desert Storm." *Landscape Architecture* 83, no. 10: 82–83.

Symonds, Glyn. 1993. "Re-used Landscapes." *Landscape Design* 219: 37–40.

Taira, Koji. 1972. "Urban Poverty, Rag-Pickers, and the Ants' Villa in Tokyo." In *Human Identity in the Urban Environment,* edited by Gwen Bell and Jaqueline Tyrwhitt, 605–11. Harmondsworth, U.K.: Penguin.

Tarr, Joel, and Gabriel Dupuy. 1988. "Sewerage in the United States." In *Technology and the Rise of the Networked City in Europe and America,* 159–85. Philadelphia: Temple University Press.

Tarr, Joel A. 1984. "Water and Waste: A Retrospective Assessment of Wastewater Technology in the United States, 1800–1932." *Technology and Culture* 25, no. 2: 226–63.

———. 1996. *The Search for the Ultimate Sink : Urban Pollution in Historical Perspective*. Akron, Ohio: University of Akron Press.

———. 1999. "Water and Wastewater: The Origins of the Piped Society." In *The Composting Toilet System Book*, edited by David Del Porto and Carol Steinfeld, 211–16. Concord, Mass.: The Center for Ecological Pollution Prevention.

Taylor, John. 1992. "Take Back the Water." *Landscape Architecture* 82, no. 5: 50–55.

Tekin, Latife. 1993. *Berji Kristin, Tales from Garbage Hills*. Translated by Ruth Christie and Saliha Paker. London: Marion Books.

Thayer, Robert. 1994. *Gray World, Green Heart: Technology, Nature, and the Sustainable Landscape*. New York: Wiley.

———. 1995. "Increasingly Invisible Infrastructure." *Landscape Architecture* 85, no. 6: 136.

Thompson, Michael. (1979). *Rubbish Theory: The Creation and Destruction of Value*. New York: Oxford University Press.

Thurgood, Simon. 1995. "Pond Technology." *Landscape Design* 243: 46–49.

Thurman, Patrick C. 1985. "Wastewater Treatment Plants." *Landscape Architecture* 75, no. 6: 52–54.

Tierney, John. 1996. "Recycling Is Garbage." *New York Times Magazine*, June 30, C1.

Todd, John, and Nancy Jack Todd. 1994. *Eco-Cities to Living Machines: Principles of Ecological Design*. Berkeley, Calif.: North Atlantic Books.

Tompkins, Judith, L. 1984. "Swan Song for a Dump in Bellevue, Michigan." *Small Town* 14, no. 2: 4–8.

Turner, Daniel. 1999. "America's Crumbling Infrastructure." *USA Today Magazine*,May, 10–17.

Ukeles, Mierle Laderman. 1992. "A Journey: Earth/City/Flow. *Art Journal* 51, no. 2: 12–14.

———. 1995. "Mierle Laderman Ukeles." In *Sculpting with the Environment: A Natural Dialogue*, edited by Baile Oakes, 184–93. New York: Van Nostrand Reinhold.

———. 1996. "Fresh Kills, Imaging the Landfill / Scaling the City." In *City Speculations*, edited by Patricia Phillips, 88–93. New York: Queens Museum of Art; Princeton Architectural Press.

Vaccarino, Rossana. 1995. "Hargreaves Associated—Remade Landscapes." *Lotus International* 87: 83–93.

Vaux, Calvert. 1857. *Villas and Cottages*. New York: Harper and Brothers.

Vigarello, Georges. 1988. *Concepts of Cleanliness: Changing Attitudes in France since the Middle Ages.* Translated by Jean Birrell. New York: Cambridge University Press.

Walker, Jesse, and Pierre Desrochers. 1999. "Recycling Is an Economic Activity, Not a Moral Imperative." *The American Enterprise* 10, no. 1: 74–75.

Walker, Peter, and William Johnson and Partners. 1994. "Countryside Landfill." *Process Architecture* 18: 52–53.

Wastewise Watch. 1998. *Wastewise Watch* (Georgetown, Ontario) 16 (March).

Weintraub, Irwin. 1994. "Fighting Environmental Racism: A Selected Annotated Bibliography." *Electronic Green Journal* no. 1 (June). Accessed from http://egj.lib.uidaho.edu/egj01/weint01.html (4/19/03).

Weisberg, Gabriel. 1993. "Scatological Art." *Art Journal* 52, no. 3: 18–19.

Weyl, Martin, ed. 1999. *Hiriya in the Museum: Artists' and Architects' Proposals for Rehabilitation of the Site.* Exhibition catalog. Tel Aviv: Tel Aviv Museum of Art.

Williams, Rosalind. 1990. *Notes on the Underground: An Essay on Technology, Society, and the Imagination.* Cambridge Mass.: MIT Press.

Wilson, Forrest. 1993. "Firmness Commodity and Delight in the Junkyard." *Blueprints* 11, no. 3: 2–6.

Winslett Evans, Rachel. 1996. "The Creation of Byxbee Landfill Park." *Critiques of Built Works of Landscape Architecture.* Vol. 3, 16–20. Baton Rouge: LSU School of Landscape Architecture.

Wright, Lawrence. 1960. *Clean and Decent: The Fascinating Story of the Bathroom and Water Closet.* New York: Viking.

Yeomans, Alfred. 1916. *City Residential Land Development.* Chicago: University of Chicago Press.

Young, Dennis. 1972. *How Shall We Collect Garbage?* Washington, D.C.: Urban Institute.

Young, William. 2001. "Fresh Kills Landfill: The Restoration of Landfills and Root Penetration." In *Manufactured Sites: Rethinking the Post-industrial Landscape,* edited by Niall Kirkwood, 178–90. New York: Spoon Press.

Yúdice, George. 1988. "Marginality and the Ethics of Survival." *Universal Abandon? The Politics of Postmodernism,* edited by Andrew Ross, 214–36. Minneapolis: University of Minnesota Press.

Yung, Susan. 1996. "Mierle Laderman Ukeles: All Systems Flow." *Appearances* 23, no. 2: 26–29.

Zavarella, Mario. 1983. "Designing a Connecticut Landfill for Reuse Potential." *Small Town* 13, no. 1: 14–17.

Index

About the Author

Mira Engler is an associate professor in the Department of Landscape Architecture in the College of Design at Iowa State University. She is a graduate of the B.L.A. program at the Technion, the Israeli Institute of Technology in Haifa, Israel, and the M.L.A. program at the University of California at Berkeley. Engler's areas of interest are marginal and waste sites, landscape art, public and environmental art, and the cultural landscape. She has published articles in *Landscape, Landscape Journal, Places, Land Forum,* and *Public Art Review.* She created a number of public art installations in Iowa and Missouri and won national awards for her conceptual designs "Rural Reliquaries" in the 1995 Visionary Landscape competition and "Open Waste System Parks," with Gina Crandell, in the 1991 Once and Future Park competition. Her conceptual proposal for the infamous garbage mountain, Hiriya, near Tel Aviv, Israel, was part of a group exhibition titled *Hiriya in the Museum* at the Tel Aviv Museum of Art in 1999–2000.

A native of Holon, Israel, Engler now resides in Ames, Iowa.

Other Books in the Series

Ecological Planning: A Historical and Comparative Synthesis
    Forster Ndubisi

Big Plans: The Allure and Folly of Urban Design
    Kenneth Kolson

From Garden City to Green City: The Legacy of Ebenezer Howard
    Edited by Kermit C. Parsons and David Schuyler

Preserving Cultural Landscapes in America
    Edited by Arnold R. Alanen and Robert Z. Melnick,
    with a foreword by Dolores Hayden

Restoring Women's History through Historic Preservation
    Edited by Gail Lee Dubrow and Jennifer B. Goodman

Waterstained Landscapes: Seeing and Shaping Regionally Distinctive Places
    Joan Woodward